ROUTLEDGE LIBRARY EDITIONS:
FOOD SUPPLY AND POLICY

Volume 7

UNITED STATES FOREIGN ECONOMIC POLICY-MAKING

UNITED STATES FOREIGN ECONOMIC POLICY-MAKING

An Analysis of the Use of Food Resources 1972–1980

KENNETH A. GOLD

LONDON AND NEW YORK

First published in 1987 by Garland

This edition first published in 2020
by Routledge
2 Park Square, Milton Park, Abingdon, Oxon OX14 4RN

and by Routledge
52 Vanderbilt Avenue, New York, NY 10017

Routledge is an imprint of the Taylor & Francis Group, an informa business

British Library Cataloguing in Publication Data
A catalogue record for this book is available from the British Library

ISBN: 978-0-367-26640-0 (Set)
ISBN: 978-0-429-29433-4 (Set) (ebk)
ISBN: 978-0-367-27578-5 (Volume 7) (hbk)
ISBN: 978-0-367-27594-5 (Volume 7) (pbk)
ISBN: 978-0-429-29676-5 (Volume 7) (ebk)

Publisher's Note
The publisher has gone to great lengths to ensure the quality of this reprint but points out that some imperfections in the original copies may be apparent.

Disclaimer
The publisher has made every effort to trace copyright holders and would welcome correspondence from those they have been unable to trace.

FOREIGN ECONOMIC POLICY OF THE UNITED STATES

United States Foreign Economic Policy–Making:

An Analysis of the Use Of Food Resources

1972–1980

KENNETH A. GOLD

GARLAND PUBLISHING, INC.
NEW YORK & LONDON • 1987

For a complete list of the titles in this series, see the final pages of this volume.

Library of Congress Cataloging-in-Publication Data

Gold, Kenneth A. (Kenneth Alan)
United States foreign economic policy-making.

(Foreign economic policy of the United States)
Originally presented as the author's thesis (Ph.D.--Pennsylvania State University, 1982)
Bibliography: p.
1. Produce trade--Government policy--United States. 2. Agriculture and state--United States. 3. United States--Commercial policy. 4. United States--Foreign economic relations. I. Title. II. Series.
HD9006.G64 1987 382'.456413'00973 87-23785
ISBN 0-8240-8083-1

All volumes in this series are printed on acid-free, 250-year-life paper.

Printed in the United States of America

TABLE OF CONTENTS

CHAPTER 1

AN ANALYSIS OF UNITED STATES FOREIGN AGRICULTURAL POLICY

Since the late 1940s, American farmers have consistently produced more food than could be consumed within the United States. Ever since that time, American decision-makers have sought to utilize those surplus agricultural commodities for purposes of supporting United States foreign policy. At the same time, other decision-makers have attempted to use the overproduction of food resources for the benefit of domestic economic interests.

United States foreign agricultural policy, like international economic policy of which it is part, has traditionally been conceived as being primarily the product of two, often competing sets of interests: domestic economic policy and general foreign policy. Domestic economic objectives included supporting the income of American farmers; insuring domestic consumers adequate supplies of food at reasonable prices; promoting growth of the economy and limited redistribution of income; and supporting the economic interests of American agribusiness corporations. The most important representative of these interests within the government was the Department of Agriculture.

Foreign agricultural policy has also been utilized in support of an extremely wide range of foreign policy objectives, ranging from strategic and ideological interests to humanitarian assistance. Begining in the early 1950s and continuing until the early 1970s the primary vehicle for utilizing agricultural resources in support of foreign policy objectives was the United States food aid program. Following the so-called "world food crisis" of 1972-1974, the international food system underwent drastic change and the food aid program was virtually eliminated. The vast majority of U. S. food exports since that time have taken place in the commercial market. Nevertheless, the United States continues to utilize its agricultural resources for purposes of foreign policy. These "state" interests are represented primarily by the White House and Department of State.

This study will attempt to develop an analytical framework for understanding United States foreign agricultural policy through a "state interest" approach, and to describe and analyze seven cases of U. S. food policy decisions through this perspective. In chapter one, a state interest approach will be constructed, and it will be demonstrated why such an approach may be superior to alternative Marxist and pluralist frameworks. Chapter one will also introduce the changing circumstances in the formulation of United States foreign agricultural policy, and in particular those alterations that occurred beginning in 1972. Chapter two offers an analysis of the nature of foreign economic policy, and outlines the reemergence of economics as an important component of U. S. foreign policy. Chapter three offers a description and analysis of the concept of "food power," including an explanation of how agricultural resources may be utilized for national purposes. Chapters four through seven present seven case studies in the formulation of United States foreign policy. Chapter eight presents summary analyses of the case studies, and the conclusions of the study.

The primary analytical foundation of both a statist and neo-mercantilist approach is the assumption that the primary determinant of policy is the interst of the state. According to a statist approach,[1] certain sectors of the government consistently pursue goals and objectives which can be most usefully understood as representing the general interest of the state in international politics, rather than as the outcome of organizational dynamics, or as the particularistic interests of groups within society. Such state goals and objectives constitute what is traditionally termed the "national interest." A statist approach should not, however, be confused with a realist framework of international politics, which attributes the behavior of a state to the pursuit of a unified national interest, and largely ignores the process of policy formulation.

No similar kind of assumption is made here, as a fundamental aspect of a statist approach is the recognition that those sectors representing the interest of the state will inevitably confront other actors in the policy process who are likely to be pursuing conflicting objectives. This approach differs as well from the current mainstream approaches to the analysis of U. S. foreign policy, liberal and bureaucratic politics perspectives, as an integral component of a statist approach is the idea that it is useful to conceive of the state as an autonomous actor, a notion which is rejected by liberal as well as Marxist perspectives.

The primary theoretical assumption of "neo-mercantilism"[2] is the partial subordination of the economy to the national interest of the state.[3] As in a statist approach, the state is regarded as an autonomous or "organic" unit which determines the national interest. According to neo-mercantilist thought, the primary objctive of foreign economic policy, like foreign policy in general, is the maintenance of the nation's security and economic well-being. A neo-mercantilist approach thus differs in many important respects from both liberal and Marxist perspectives of international political economy.[4]

A major purpose of this study will be to demonstrate that a state interest approach is superior to alternative analytical frameworks for understanding the process of United States foreign agricultural policy-making. A series of case studies will provide evidence that American central decision-makers since the early 1970s have sought on most occasions to utilize U. S. food resources to accomplish foreign-policy objectives which can be most usefully understood in terms of the interests of the state.

Understanding Foreign Policy Through a Statist Perspective

The basic analytical assumption of a statist approach to the study of foreign policy is that there exists a distinction between the state and society. This approach views the state as an autonomous actor. According to Stephen Krasner:

> This approach makes a critical assumption: that it is useful to conceive of a state as a set of roles and institutions having peculiar drives, compulsions, and aims of their own that are separate and distinct from the interests of any particular societal group. These goals are associated either with general material objectives or with ambitious ideological goals related to beliefs about how societies should be ordered. They can be labelled the national interest. In striving to further the national interest, the state may confront internal as well as external resistance. Central decision-makers may be frustrated not only by other states but also by their inability to overcome resistance from within their own society.[5]

Thus, according to a statist approach, the state is seen as formulating goals autonomously, which it then attempts to implement in a setting which may offer resistance both domestically and internationally.

In United States foreign policy, the national interest is represented primarily by the White House and the Department of State, as evidenced in cases where discrete decisions can be identified, and isolated analytically. Such a conceptualization would apply more consistently to "major" decisions, rather than the incremental policy process which goes on within the various departments and agencies.

> What distinguishes these roles and agencies is their high degree of insulation from specific societal pressures and a set of formal and informal obligations that charge them with furthering the nation's general interests . . . In the United States, at least in the realm of foreign affairs, the White House and the State Department are the pivot of the state.[6]

A statist approach does not try to suggest that the nation's interests are unrelated to society's interests. Neither, however, is it merely a "summation" of the interests of particular groups or individuals. Here Krasner cites Pareto's work which draws a distinction between the "utility _of_ the community" and the utility _for_ the community".

> Utility _for_ the community involves summing the preferences of individual members of a community. The utility _of_ the community involves making a judgment about the well-being _of_ the community as a whole.[7]

According to this perspective, then, the state determines a set of values according to its judgment regarding the utility of the community. Based on these values the state identifies certain general objectives, which become the basis of the national interest. This does not imply, however, as a realist paradigm does, that such objectives somehow automatically become the foreign policy doctrine of the United States.

Rather, a statist approach suggests that it is useful to associate these sorts of general objectives with certain sectors of the government which consistently act as representatives of the state in the foreign policy process. Unlike a realist paradigm, the term "state" is used to refer to only certain sectors of the political system, rather than some undefined whole. And unlike a realist paradigm, which is intended as a framework for understanding the behavior of states in international poltics, Krasner's statist approach is formulated as a framework for understanding the formulation of foreign policy:

> This is not a study of billiard balls, of the way in which individual countries treated as unified wholes interact in the international system. On the contrary, the approach used here explicitily recognizes the need to examine the policy-making process within a country when dealing with questions involving foreign policy. A state must deal with private actors in its own society as well as with other actors in the international arena.[8]

The realist perspective portrays the government as a rational actor or black box. A statist approach, however, looks inside the black box in the form of posing two central problems for analysis: identification of the major objectives of central policy-makers; and analysis of policy-makers' ability to accomplish these objectives within society as well as in the international system.

Writers of the realist school do not generally concern themselves with examining the policy process which takes place in the formation of the national interest, but rather treat the concept of national interest as a basic assumption. Hans Morgenthau, for example, states that the national interest is comprised of a set of general principles which guide a state in its relations with other states. He defines the "primary" national interest as the safeguarding of a state's territory and institutions. Based on this assumption, propositions about how a state will behave in the international system are arrived at deductively. Krasner attempts to use the concept of national interest in a way which differs significantly with the realist approach. He states:

> Here the national interest is defined inductively as the preferences of American central decision-makers. Such a set of objectives must be related to general societal goals, persist over time, and have a consistent ranking of importance in order to justify using the term "national interest". For any issue it is not difficult to make a list of aims desired by political leaders. These range from satisfying psychological needs to increasing wealth, weakening opponents, capturing territory, and establishing justice.[9]

It will be shown that since the late 1940s the foreign agricultural policy of the United States has openly and consistently served as an adjunct to its general foreign policy. In broadest possible terms, this has meant the maintenance of national security against all threats, whether real or perceived, from the outside. While foreign policy resources have been utilized in support of secondary objectives such as for commercial or humanitarian reasons, the major theme of United States foreign policy since World War II has unquestionably been anti-Communism. This is, of course, not meant to imply that all actions of the United States government in the postwar period have been directed toward this single objective. What it does mean is that certain major "shared images" are held by central decision-makers, and that these images provide a common denominator which serves as a basis for policy preferences. Morton Halperin has identified a list of the shared images of the postwar period which is widely subscribed to:

> 1. The preeminent feature of international politics is the conflict between Communism and the Free World.
> 2. Every nation that falls to Communism increases the power of the Communist bloc in its struggle with the Free World.
> 3. The surest simple guide to U. S. interests in foreign policy is opposition to Communism . . .[10]

The problem for analysis is to be able to demonstrate that this primary objective remains paramount in the preferences of central decision-makers over time. If this cannot be accomplished then it would not be possible to specify a national interest by induction. According to Krasner's analysis, the national interest defined in terms of the promotion of broad ideological goals, he asserts, supports the thesis that American central decision-makers were more concerned with structuring the world according to a vision based on American values than in furthering specific economic and strategic interests. This was made possible as a result of America's preeminent position of power in the international arena. Krasner states:

> By the end of the Second World War, American central decision-makers commanded a set of power resources unprecedented in modern times. These resources allowed them to turn toward projecting America's vision of a properly ordered society into the international system. They were freed from specific strategic and economic concerns. The distribution of power in the international system is the critical variable in determining the broad foreign policy goals sought by American central decision-makers.[11]

Moreover, the manner in which such ideological objectives were pursued in American policy is crucial to distinguishing Krasner's statist perspective from a Marxist position. The fact that central decision-makers often formulated policy in a "nonlogical" manner, by persistently misperceiving the communist threat, and by miscalculating America's ability to affect the course of events almmost anywhere in the world, Krasner claims, offers strong evidence in support of a statist paradigm. A Marxist position, which ascribes foreign policy behavior to the influence of particular groups in society, or which portrays the state as acting to preserve capitalist society, cannot readily accommodate the pattern of United States foreign policy since the Second World War:

> In their pursuit of an ideological foreign policy U. S. leaders undermined the coherence of American society: and it is difficult to associate the actions of American policy-makers, particularly in Vietnam, with contradictory pressures emanating from capitalist societal structures. On the contrary, it was the ability of American leaders to free themselves from societal constraints (reflecting America's hegemonic position in the international system) that allowed them to define and pursue ideological goals in a nonlogical manner, even though this activity weakened the fabric of American society.[12]

Franz Schurmann supports the thesis that important aspects of United States foreign policy since the end of the Second World War can be most usefully understood in terms of the pursuit of an ideological "vision". No other interpretation can satisfactorily explain American policies which attempted to bring peace and prosperity to the world. Furthermore, Schurmann maintains, that vision was promoted against domestic opposition by powerful sectors of the American business class who were "traditionally the most expansionist segment of American society."

> This points to something that people raised in or influenced by the Marxist tradition are prone to reject: the autonomous, innovative, and powerful role of the state . . .
>
> The ideological currents in America that relate to its foreign policy from the end of World War II to the present do not relate directly either to the material interests of people or to the social quality of life. They relate to matters of order, security, and justice. When the United States became an empire at the close of World War II, foreign affairs assumed a very large place in the political minds of Americans. The ideologies propounded to justify the empire did not promise the people booty from foreign exploitation, nor did they suggest that the quality of neighborhood life would be improved by foreign aid to Nepal. Ideological appeals were put entirely in terms of the dire consequences for America of world chaos, the threats from foreign aggressors, and the chance to implement American ideals in the benighted parts of the world.[13]

An ideologically based foreign policy can be understood best from a statist perspective because no specific groups in society would be expected to asociate their particularistic interests with an ideological vision, even though some sets of interests, such as certain corporate economic objectives, may at times benefit. Such a policy could only be formulated by central decision-makers sufficiently insulated from societal pressures so as to be able to pursue aims which impose costs on powerful nonstate actors, or even society as a whole.

Krasner believes that a further understanding of a statist perspective can be demonstrated by contrasting it with two major alternative approaches to the political process, Marxism and liberalism.

Marxism

Krasner utilizes a distinction subscribed to by a number of analysts which divides Marxist theories of the state into two major types: instrumental Marxism and structural Marxism. While Marxists in general present the most extensive analysis of foreign policy, he asserts that of the two types, it is more difficult to distinguish a structural Marxist approach from a statist perspective. Instrumentalist theories are characterized by a systematic examination of the ties between the ruling class and the state. As interpreted by Gold, Lo, and Wright:

> The instrumentalist perspective provides a fairly straighforward answer to the question, 'Why does the state serve the interests of the capitalist class?' It does so because it is controlled by the capitalist class . . .
>
> The functioning of the state is . . .fundamentally understood in terms of the instrumental exercise of power by people in strategic positions, either directly through the manipulation of state policies or indirectly through the exercise of pressure on the state.[14]

A structuralist theory of the state differs from the instrumentalist perspective in that it rejects the idea that the state serves as a mere "instrument" of the ruling class. Instead, the structural analysis focuses on the structures of society which are said to determine broadly the functions of the state, rather than on individuals who may occupy positions of power. Gold, Lo, and Wright state:

> The structuralist theory of the state . . . attempts to unravel the functions the state must perform in order to reproduce capitalist society as a whole. These functions determine the specific policies and organization of the state.[15]

Krasner takes a similar position:

> Structuralist Marxist arguments . . . do not attempt to trace the behavior of the state to the influence of particular capitalists or the capitalist class. Instead, they see the state playing an independent role within the overall structure of a capitalist system. Its task is to maintain the cohesion of the system as a whole. At particular times this may require adopting policies opposed by the bourgeoisie, but generally official action and the preferences of leading elements in the capitalist class will coincide.
>
> For structural Marxists, the state can be treated as having autonomy, not from the needs of the system as a whole, but from direct political pressure from the capitalist class.[16]

It will not be within the scope of this paper to seek to determine the merits of the distinction between the structural and instrumental Marxist positions. It may be pointed out, for example, that Nicos Poulantzas, often described as a leading exponent of the structuralist perspective, rejects such a distinction. All Marxist theories of the state do, however, share a basic proposition.

> . . . all Marxist treatments of the state begin with the fundamental observation that the state in capitalist society broadly serves the interests of the capitalist class . . .
>
> Given this axiom, Marxist theories of the state generally attempt to answer two complemntary questions: 'Why does the state serve the interests of the capitalist class?' and 'How does the state function to maintain and expand the capitalist system?'[17]

Krasner believes that the state-centrist apporach differs from the analytic assumptions of Marxist theories of the state in at least three important ways:

> First, the notion of national-interest is rejected by Marxists. The aims pursued by the state mirror the preferences of the capitalist class or some of its elements, or the needs of the system as a whole. State behavior does not reflect either autonomous power drives or the general well-being of the society. Second, the behavior of the state is taken by them to be intimately related to economic goals: other objectives are instrumental, not consummatory. In particular, ideological objectives cannot be independent of economic considerations . . . Third, even though structural Marxists may view the state as relatively autonomous, they do not believe that it can really be understood outside of its societal context. The state has peculiar tasks within the structure of a capitalist system, but they are ultimately associated with the interests of a particular class.[18]

Richard Barnet and Ronald Muller take a similar position in their analysis of global corporations, believing that the interests of the state and corporations have become increasingly blurred to the point where distinction between the two is now obsolete. They cite Karl Polanyi's argument which traces the use of the "market mentality" in the laissez-faire atmosphere of the Industrial Revolution as the basis for transformation of the relationship of private and public government. With the creation of this new type of "economic society", economic goals measured by profits and product became the primary determinant of the functions of both corporations and the state.

Thus, economics became "the substance of politics," as both the state and corporations grew and expanded their power together. According to this view, it is not so much that corporations determine the national interest, but that basic goals such as growth and stability are shared by both. This definition, according to Barnet and Muller, is subscribed to by both Marxists and capitalists:

> For the Marxist, the government is the instrument of the dominant economic power in the society. Since the state is the 'executive committee of the ruling class,' to use Marx's phrase, it is inconceivable that the interests of the corporations could clash with the interests of the state. At the same time, it is accepted capitalist dogma that what is good for corporations is good for the United States and vice versa. The richer and more powerful global corporations became, the more these riches and power redound to the benefit of the United States.[19]

A Marxist position, then, defines the interest of the state as a reflection of the interests of the capitalist class, or corporation. The idea of a state interest defined in ideological terms is therefore incompatible. In many areas of foreign economic policy, the corporations and the state have shared a complementary set of interests. In a number of cases involving United States foreign agricultural policy, however, it will be shown that the interests of many corporations were in opposition to the interests of the state, and in many of the cases the interest of the state prevailed.

Liberalism

Krasner also contrasts a statist perspective with liberal, or interest-group approaches to foreign policy analysis. In its most general form, liberalism views the political process as a forum of competition among a variety of organized interests. The government policy which emerges as a result of this process reflects a complex series of compromises from among those pressures which are most effectively brought to bear upon governmental institutions. Arthur F. Bentley first introduced the term interest, or group interest, in his 1908 publication The Process of Government. Since that time, the concept of the group as the basic political form has been dominant among observers of American politics. In recent years, the logic of interest group theory has been applied to the analysis of United States foreign policy formulation in the form of the beaucratic politics paradigm. While a statist perspective shares much of its intellectual underpinning with the bureaucratic politics approach, the differences between the two are significant.[20]

According to its proponents, the bureaucratic politics paradigm was designed essentially as an alternative framework for explaining foreign policy decisions to the previousl pervasive realist or rational actor model. Morton Halpern has summarized the classical model:

> In trying to explain foreign policy decisions, most observers assume that decision makers are motivated by a single set of national security images and foreign policy goals. Supposedly decisions reflect these goals alone, and actions are presumed to flow directly from the decisions. Thus 'explanation' consists of identifying the interests of the nation as seen by the leaders and showing how they determine the decisions and actions of the government.[21]

Advocates of the bureaucratic politics paradigm, however, believe that "the reality is different". According to their model, policy is viewed as the outcome of a process in which various groups compete and bargain with each other. According to Halpern:

> The actions of the American government related to foreign policy result from the interests and behavior of many different groups and individuals in American society. Domestic politics in the United States, pubic attitudes, and the international environment all help to shape decisions and actions. Senators, congressmen, and interest groups are involved to varying degrees, depending on the issue. The relevant departments of the federal bureaucracy are involved, as is the President, at least on major issues.[22]

The bureaucratic politics paradigm suffers from a number of serious shortcomings. First, it confuses levels of analysis, a problem which appears endemic to foreign policy analysis. The bureaucratic politics paradigm, as formulated by Graham Allison and others, was originally intended as an alternative to the classical or rational actor model, with which it found fault for ignoring the process of policy formulation. The classical model, however, is an international politics framework, which attempts to explain why states behave as they do in the international system. The disadvantages of the adoption of a macro, or systemic level of analysis have been recognized for some time. As J. David Singer has observed:

> By eschewing any empirical concern with the domestic and internal variations within the separate nations, the system-oriented approach tends to produce a sort of 'black box' or 'billiard ball' concept of the national actors.

Yet the systemic level of analysis, while requiring the attribution of a greater degree of similarity to national actors than obviously exists in reality, also contains certain positive attributes as an analytical model:

> The systemic level of analysis, and only this level, permits us to examine international relations in the whole, with a comprehensiveness that is of necessity lost when our focus is shifted to a lower, and more partial, level.[23]

Second, the bureaucratic politics model underestimates the influence of the Executive in foreign policy formulation. A statist perspective does not seek to contradict the idea that governmental policy can usefully be understood in terms of the outcome of bargaining among the various parts of the bureaucracy. It does, however, reject the idea of a framework which fails to recognize the special role of the presidency in the formulation of foreign policy. As John Spanier and Eric Uslander observe:

> The central and key figure in the foreign policy process is the President. Although noncrisis policy is largely the product of bureaucratic conflict, this does not mean--as models of bureaucratic struggle often imply - that the President is merely one player among many, one chief among many departmental chiefs. He

> is more than merely a first among equals; he is by far the most important player.[24]

This is not to suggest that theorists of the bureaucratic politics school do not recognize the special position of the President as leader and preeminent decision-maker in United States foreign policy. Writers of the school generally subscribe to the dictum, first advanced by Richard Neustadt, that the power of the President is basically the power to influence and persuade.[25] This perspective also acknowledges that is is the President who usually makes the final decision. But the bureaucratic politics school explains the position and actions of the President in terms of inputs from the bureaucracy. According to this analysis, the President's stand on any particular issue depends largely on which participants from the bureaucracy he selects on the basis of regularized channels of access. A statist perspective, on the other hand, asserts that in addition to commanding a preeminent position in the policy-making structure, the nature of the President's policy objectives is different from other sectors of government, as only the chief executive can represent the national interest.

The third deficiency of the bureaucratic politics paradigm is that it fails to consider the concept of national interest as based on the analytical assumption that there exists a distinction between the state and society. The rejection of the idea of a conceptually distinct state is a basic caveat of pluralist theory. Theodore Lowi agrees that this is a fundamental problem of contremporary pluralism:

> The zeal of pluralism for the group and its belief in an natural harmony of group competition tended to break down the very ethic of government by reducing the essential conception of government to nothing more than another set of mere interest groups . . .
>
> It should thus be evident that pluralist theory today militates against the idea of separate government. Separate government

> violates the basic principle of the automatic political society. This was reinforced by the scientific pluralist's scientific dread of such poetic terms as public interest, the state, and sovereignty that admittedly cannot be precisely defined and are closely associated aesthetically with the notion of separate government. But by such means pluralism gained a little and lost a lot.[26]

If government policy is viewed primarily as a reflection of the interests and goals of particular groups in society, or particular organizations in government no concept such as national interest can exist as well. The concept of national interest is based on a distinctly alternative notion, that certain policies can most usefully be explained as the result of the state conceived as an autonomous actor pursuing its own peculiar interests, such as broad material or ideological objectives. According to Krasner's analysis:

> Analysts adopting a liberal or pluralist perspective have been very critical of the concept of the national interest. An inescapable implication of their position is that government policy is a reflection of whatever groups have power in the society. The concept of the public interest slips away . . . .
>
> Liberal conceptions of politics also have little use for the notion of the state as an autonomous actor motivated by drives associated with its own need for power or with the well-being of the society as a whole. Governmental institutions merely process inputs and outputs. The state is seen as a set of formal structures, not an autonomous actor.[27]

Defining the National Interest

Utilization of a statist approach for the analysis of foreign policy formulation must begin with the identification of the national interest, defined as the goals and objectives of central decision-makers. Krasner distinguishes two approaches to studying the national intererst. The first approach is the "logical-deductive" formula, characterized by the realist paradigm in international relations analysis, which is based on the assumptions concerning the behavior of the state.[28] It is assumed, for example, that the state will pursue certain objectives dictated by defining the national interest in terms of defending certain irreducible interests. This assumption then serves as a basis for formulating certain propositions about how a state will be expected to behave in a given situation. Although a realist approach may be useful in helping to predict certain forms of state behavior in the international system, it suffers from two major drawbacks for foreign policy analyis. First, it is primarily suited for addressing major issues, or "high policy" questions. These are questions connected with defending core objectives perceived by central decision-makers as threats to the security of the state's territory and institutions. Many important foreign policy issues, such as those related to international economic policy, cannot always be accounted for in these terms. Second, many important insights which might be gained from examining the process of formation of the national interest are ignored by the logical deductive formula.

Krasner suggests that in utilizing a statist perspective, a more useful way of defining the national interest would be through an "empirical-inductive" approach. By this method, the national interest is inducted by examining the statements and preferences of central decision-makers. Such preferences can be termed the national interest if they satisfy two basic conditions: first, they must be capable of being interpreted as reflecting the general interest of society, more than as the preferences of particular groups, or motives of individual decision-makers; second, the priority of objectives must remain consistent over time. An important function of this alternative approach to defining the national interest is to be able to apply the concept to issues not directly related to the state's core objectives. Krasner states:

> Although the logical-deductive approach may be very powerful under certain circumstances, it must be supplemented by an alternative statist interpretation of the national interest--one that is inductive and empirical rather than *a priori*. A second way of apprehending the national interest is to examine what state actors say and do. In some ways this is the most straightforward way of dealing with the concept.[29]

Implementing State Objectives

As already stated, an integral characteristic of a statist approach is the recognition that the state, in pursuing the goals and objectives associated with the national interest, is likely to encounter internal, as well as external resistance to its policy preferences. Unlike the realist paradigm in international politics, the statist approach emphasizes the internal foreign policy-making process. According to Krasner's analysis, the major variable which determines the ability of the state to implement successfully its policy preferences in the face of domestic opposition is the strength of the state in relationship to its own society.

Krasner attempts to classify the strength of the state in society by posing a continuum ranging from weak to strong. At the weak end one finds a state completely controlled by societal interest groups. The status of central government institutions is near disintegration, completely at the mercy of specific pressure groups. He cites Lebanon before the 1975 civil war as an example of this unusual circumstance. At the other end of the spectrum is the condition whereby a state is able to dominate society completely. Here, the state possesses the capabilitiy to alter the structure of social and economic institutions. Such states have historically existed immediately following major revolutions. In such situations, it is really the extreme weakness of society which abdicates its functions to the state, rather than some sudden aggrandizement of power by the state itself. The Soviet Union after 1917 and China after 1949 are examples of this condition.

The strength of most states, of course, falls somewhere in between these two extremes. Krasner has devised three ideal types for categorizing the strength of the state in capitalist or market economy societies, where some degree of autonomy can be said to exist between public and private institutions. First, in a "weak" pattern of state power, the state is not so strong as to be able to alter the behavior of private groups, but may be capable of resisting pressures from such groups. Here, the state is incapable of directing the behavior of large corporations toward achieving the interests of the state, or create alternative institutions such as state-owned corporations, but may be able to ignore corporate preference.

Second, in a moderate pattern of state power vis-a-vis its domestic society, the state is not only capable of resisting private pressures, but also possesses sufficient strength so as to be capable of persuading private actors to adopt policies in support of state interests. In such a system, however, the state is not so strong as to be capable of imposing structural changes on society, and so it must work within existing societal structures.

Third, in a "strong" pattern of state power, the state possesses not only sufficient strength to alter the behavior of private groups but, in time, may be able to alter existing social structures as well. Here the state has a broad spectrum of instruments by which it can direct society in support of its policy preferences. It could, for example, create new economic structures, assume control of corporations, or grant favors to those sectors of the economy with which it shared interests. Each of these patterns of state power in relationship to society is incompatible with either a liberal or Marxist theory of the state. Although the strength of the state varies considerably between a weak pattern and strong pattern of state power, these categories are all based on some degree of autonomy of the state from its own society.

In general, developed capitalist societies such as the United States will most often exhibit a "weak" or "moderate" pattern of state power. The American political system is regarded as possessing fragmentation and dispersion of power and authority. The concept of dividing power, as well as the concept of limited central government all act to provide multiple points of entry to the political process. Krasner believes that this central characteristic of the American political system, the condition of a "strong society but a weak state," is a major determinant of the ability of the state to implement the national interest:

> The American state--the President and those bureaus relatively insulated from societal pressures, which are the only institutions capable of formulating the national interest--must always struggle against an inherent tendency for power and control to be dissipated and dispersed. They must operate in a political culture that views the activist state with great suspicion. This is particularly true of the business sector. American capitalists have a more negative reaction to public economic activity than their counterparts in other advanced market economies . . . . In trying to promote the national interest, the American state often confronts dissident bureaus, a recalcitrant Congress, and powerful private actors.[30]

The power of the state, and therefore its ability to implement its policy preferences will vary significantly, however, depending on the particular issue area involved. Certain core areas of national security policy, for example, are generally regarded as being relatively immune from domestic opposition, although even this realm has been subject to increasing challenge in recent years. And although particular policies and methods may be and often are questioned, a basic area of shared interests has generally predominated in national security issues.

Two important qualifications to this condition must be considered with regard to the present era. First, in recent years opposition has grown not only to the methods by which national security interests are pursued, but to the area of core interests itself. The American Congress in particular has played an increasingly prominent role in affecting the ability of central decision-makers to implement their objectives. Second, foreign economic issues, which in recent years have assumed an extremely important position in overall United States foreign policy, differ in certain major ways from strategic issues. As Krasner points out:

> There is, however, no reason to assume that foreign economic policy-making is identical with foreign political policy-making. Any economic decision is likely to affect groups within the society differentially, creating the potential for societal conflict. For this reason it is questionable to assume that policy can be understood solely by examining the motivations and perceptions of central decision-makers. In a political system where state power is weak and fragmented, foreign as well as domestic economic policy can be influenced or even determined by societal groups.[31]

In this sense foreign economic policy is viewed by both domestic economic interests as well as central decision-makers attempting to implement state interests as a means for achieving their respective objectives. By definition, international economic policy straddles two policy arenas: domestic economic policy and general foreign policy, or national security policy. Depending on the circumstances surrounding a particular issue, the state may determine whether to assign priority to either external diplomatic and security considerations, or to domestic economic preferences. And although in certain situations these sets of interests converge, they will more often be found in opposition to one another. Stephen Cohen offers a similar analysis:

> This duality of character means that international economic policy can be co-opted by states to promote either domestic priorities or international political and economic priorities. The problem is that both priorities normally cannot be successfully pursued simultaneously in international economic relations.[32]

The idea that the American state is relatively weak in relation to society does not of course mean that central decision-makers' preferences are never translated into policy. The most obvious case is when little or no internal opposition exists to state objectives. In the case of foreign agricultural policy in the 1950s and 1960s, Public Law 480 for the most part served the goals of both domestic economic interests and general foreign policy.

In cases where there does exist some form of domestic opposition to state preferences, Krasner suggests that two related variables act to facilitate the ability of central decision-makers to implement the national interest: the ability of the state to exercise political leadership; and the location of the decision-making arena.

It is widely recognized that a principal source of presidential power is the ability to persuade other decision-makers that a policy he wishes to pursue is in the national interest. Quite apart from the question as to whether the concept of national interest has any utility as an analytical tool, most observers would not question its value as an instrument of political action. According to an interpretation by James Rosenau, the outstanding characteristic of the concept in both uses is in its value orientation. The claim of national interest is probably most often cited by certain central decision-makers as a means of legitimizing a particular course of action to society by the assertion of "what is best" for the state. For this reason, Rosenau believes, its utility as an analytic tool is quite limited. Its real value is as a guide to political action, as well as a means of mobilizing public support. According to this analysis:

> Not only do political actors tend to perceive and discuss their goals in terms of the national interest, but they are also inclined to claim that their goals <u>are</u> the national interest, a claim that often arouses the support necessary to move toward a realization of the goals. Consequently, even though it has lost some of its early appeal as an analytic tool, the national interest enjoys considerable favor as a basis for action and has won a prominent place in the dialogue of public affairs.[33]

A similar position is subscribed to by Morton Halperin, a leading exponent of the bureaucratic politics model:

> The President stands at the center of the foreign policy process in the United States. His role and influence over decisions are qualitatively different from those of any other participant. In any foreign policy decision widely believed at the time to be important, the President will almost always be the principal figure determining the <u>general</u> direction of actions . . . .
>
> Furthermore, the President serves as the surrogate for the national interest. Many senior participants look to the President as to a blueprint for clues to the national security. His perception and judgment of what is in the national interest are dominant in the system. A strong President--with a clear sense of direction and leadership--can have a very strong influence on the images shared by bureaucratic participants, by Congress, and by the public.[34]

In addition, the degree to which central decision-makers are able to overcome societal opposition to their preferences will depend importantly on the decision-making arena. State preferences are more likely to be implemented if decision-making is located in central state institutions such s the White House or State Department. If, on the other hand, decisions are made in departments which represent the interests of particular societal groups, or in Congress which provides multiple points of access, state preferences are less likely to be implemented.

The record of United States foreign agricultural policy provides a valuable area for analysis of this thesis. During the 1970s, jurisdiction over food policy decisions appears to have oscillated between the Department of Agriculture, which managed domestic agricultural policy, and central state institutions whose primary concerns were general foreign policy.

The Formulation of United States Foreign Agricultural Policy

Prior to 1972, the Department of Agriculture was the primary initiator in United States foreign agricultural policy. As the traditional representative of domestic economic interests, the predominant concern of U. S. food policy was the welfare of the American farmer. In addition to meeting consumer demand for adequate supplies of food available at reasonable prices, the major political constituency of the Department of Agriculture was of course the farm community. Farm lobbyists have always been regarded as an effective political force, and the farm bloc vote has always been an important consideration in the formulation of food policy.

Since the end of the Second World War, the central focus of U. S. food policy has been the formulation of programs designed to support the income of the American farmer in a prevailing market situation of chronic overproduction and depressed prices. During the 1950s and 1960s, the Department of Agriculture managed massive government programs which supported commodity prices, controlled production, and increased the demand for U. S. agricultural products by promoting exports abroad.

Although these programs were also utilized for purposes of general foreign policy, state objectives did not ordinarily conflict with domestic economic interests.

The global food system which emerged after 1945 was shaped primarily by these United States policies. This unprecedented influence in international agricultural affairs was acquired as a result of America's ascendance to the world's largest exporter of grain; from its ability to determine world food prices through the grain reserve system; from its massive food aid programs; and as a corollary of the United States position of preeminent global power.

In one year, 1972, the international food system, characterized for twenty-five years by abundant agricultural surplus and low, stable prices came to what appeared to be a sudden, unexpected end. The new world food system which emerged after 1972 is far less stable than the one the world had grown accustomed to: prices have risen dramatically and are considerably more erratic; the world grain reserve has been severely cut back; and the United States food aid program, which dominated agricultural exports to the food-deficit poor countries for more than two decades, has almost completely disappeared.

Commonly cited explanations for the causes of the 1972 "world food crisis" include a decline in world food production due to unfavorable weather conditions over large areas of the world, especially the severe drought in Africa and the failure of the rice crop in Southeast Asia and Korea; a drastic decline in the Peruvian anchovy catch; the world energy crisis, and in particular the resulting high fertilizer prices; large purchases of grain by the Soviet Union; a decline in world grain stocks as a result of policy changes in the United States; and the various devaluations of the U. S. dollar, which led to increased demand for American agricultural exports.[35]

While most observers concur that most or all of these factors contributed to the abrupt alteration of the international food system following 1972, there is less agreement on whether the primary explanation of the crisis lies in a coincidence of natural occurrences, or was brought about by the successful implementation of major changes in United States foreign economic policy, which were directed toward increasing both the price and volume of American agricultural exports.

In concert with these events, fundamental changes have taken place in the process of United States foreign agricultural policy. The grain reserve system and the food aid program, which together successfully met the needs of both the domestic economic interests and foreign political objectives of United States food policy since the early 1950s, have been almost completely phased out of existence. Although central decision-makers continue to view American agricultural resources as a means of supporting the general foreign policy of the state, they must now seek to implement their objectives in an entirely new context. First, U. S. food exports now take place almost entirely in the commercial market, where they are handled by a small group of private corporations. Second, the domestic economic interests of American agricultural policy have undergone drastic change. Formerly the domain of the interests of the American farmer, and, to a lesser extent American consumers, the most influential domestic constituency of United States food policy is now corporate agribusiness and the major grain trading corporations.

Although the emergence of corporate agribusiness into the public view was quite sudden, the power and influence of these huge, diversified multinational companies have been steadily expanding for a number of years. The grain trading corporations have actually occupied a pivotal position in the international food system since the nineteenth century. Their ascension to a major role in United States food policy, however, paralleled the transformation of American agriculture. Over the last forty years, approximately 4 million family farms have gone out of existence. The farm population of the United States declined from 25 percent of the total in 1940, to 15 percent by 1950, and to 9 percent by 1960. Currently, farmers comprise less than 4 percent of the population.[36] In addition, it is reported that less than 4 percent of the farms that do remain produce more than half the total food production.[37] These developments have radically altered the political base of American agriculture.

The most influential private actors in United States foreign agricultural policy at the present time appear to be the major grain trading firms and agribusiness corporations. Five privately held multinational companies, the two largest of which are based in the United States, now control the vast majority of American agricultural exports.[38] At the top of the list is Cargill, Inc., with reported assets of more than $3.25 billion, and the Continental Grain Company. They are followed by Louis Dreyfus and Co, the Bunge Corporation, and Garnac Grain, an affiliate of the Swiss-based Andre Corporation.

The grain corporations occupy a central position in United States agricultural exports and in the world food supply system. They own or have interests in every aspect of agriculture, including local grain elevators, railroad cars, shipping barges, and port facilities. They oversee charter operations which control the shipment of agricultural commodities to all parts of the world, and own numerous port facilities and processing plants abroad. Their intellignece network gathers information on every aspect of the international agricultural trade and is so accurate that it is reported that they regularly supply information to the CIA.[39]

A major source of the grain corporations' profits is their ability to manipulate the international grain market, where they benefit from high prices, and particularly unstable prices. Their economic interest therefore lies in a free commercial trade system, unencumbered by government policies which act to stabilize the market. There is a substantial amount of evidence that the grain corporations wield considerable influence in the formulation of United States agricultural policy. According to a NACLA report:

> While advocating a free trade policy, the companies strive incessantly to influence and control the State. Key business associations, such as the National Grain Trade Council and the National Grain and Feed Association, lobby in Washington on behalf of the grain trade . . . . Be it detente with the Soviet Union or tariff negotiations with the European Common Market, the companies work closely with the U. S. government to formulate foreign policies. The power of the trade rests not only in its manipulation of the market, but in its ability to use the State for its own interests.[40]

In addition, it is widely acknowledged that since at least the early 1970s, the U. S. Department of Agriculture, formerly the guardian of the American farmer and the American consumer, has abandoned that role and now reflects the interests of the major grain trading firms and corporate agribusiness. As observed by Susan DeMarco and Susan Sechler of the Washington-based Agribusiness Accountability Project:

> Department of Agriculture officials dealing with grain have increasingly perceived their role not as regulating export operations to the benefit of both producers and consumers, but as protecting the export industry. While proper public relations deference has been paid to consumer needs and world hunger, the actual policies advocated by U.S.D.A. for foreign exports have pretty much followed those advocated by the grain trade.[41]

This study will attempt to analyze United States foreign agricultural policy-making both prior to and following these changes. Specifically, it will focus on the ability of central decision-makers to utilize agricultural resources for general foreign policy objectives. By examining a number of cases where the United States attempted to exercise "food power", it will be demonstrated that the state has consistently pursued objectives which can usefully be understood in terms of the national interest. In attempting to implement state objectives, central decision-makers have had to "share" foreign agricultural policy with domestic groups which had important economic interests in food policy.

Throughout the 1950s and 1960s, the foreign policy objectives of the state and the economic interests of the dominant domestic constituency of U. S. food policy normally converged. The major government agricultural programs of this period - Public Law 480, the grain reserve system, and the soil bank - benefited the American farmer by supporting food prices, disposing of surplus food stocks and creating increased demand for U. S. agricultural products abroad. These same programs provided the state with a valuable resource which central decision-makers were able to utilize in support of general foreign policy objectives.

This complementary interest between domestic economic groups and state foreign policy preferences was substantially altered with the appearance of a new world food system following 1972. American agricultural resources, which for more than two decades were regarded as disposable surplus, quite suddenly became a significant component of the Nixon Administration's New Economic Policy.

There are strong indications that the new U. S. position in foreign agricultural policy was part of a new philosophy concerning America's role in world affairs. The new agricultural policy can in this way be viewed as part of a retreat from U. S. positions in the world economy which had come to be regarded as part of America's responsibilities, but no longer seemed to serve U. S. interests - a sort of economic adjunct to the strategic-based Nixon Doctrine. American economic preeminence was fast becoming a relic of another era, and the Nixon Administration viewed the grain reserve system and food aid programs as policies which the United States could no long economically afford. As Secretary of Agriculture Butz declared at the time: "As we are not the world's policeman, neither are we the world's father provider."[42]

The United States had begun to experience serious balance of payments difficulties, and in 1971 registered the first trade deficit of the twentieth century. In response to the economic crisis, a presidential Commission on International Trade and Investment Strategy concluded that America's economic problems might be remedied by expanding the two areas of commercial exports where the United States maintained a competitive advantage: in high technology and agriculture. Subsequent policy decisions sought to implement a free trade doctrine in agriculture, which meant the withdrawal of government programs which had regulated farm income and productive capacity since the 1930s.

Furthermore, as the effectiveness of the traditional strategic and diplomatic instruments of American power appeared to decline, central decision-makers began to look increasingly toward economic resources as a means of supporting foreign policy objectives. The aftereffects of the Vietnam war had placed new constraints on American power, both internationally, and domestically. Third World countries were moving to assert greater control over their own resources, and the OPEC oil embargo vividly demonstrated the utility of economic weapons for political purposes. As these factors combined with the world food crisis, policy-makers gave renewed consideration to the idea of utilizing "food power" in United States foreign policy.

Chapter Notes

1. The interpretation of a statist perspective is drawn primarily from Stephen Krasner, Defending the National Interest (Princeton, N. J.: Princeton University Press, 1978).

2. The term "neo-mercantilism" is employed by various authors in order to distinguish its application from mercantilist doctrine as practiced in the seventeenth and eighteenth centuries.

3. This study will draw upon current analyses of international political economy which apply mercantilist theory to the current period. See especially Benjamin Cohen (ed.), American Foreign Economic Policy (New York: Harper & Row, 1968); Stephen D. Cohen, The Making of United States International Economic Policy (New York: Praeger, 1977); and Robert Gilpin, U. S. Power and the Multinational Corporation (New York: Basic Books, 1975).

4. Robert Gilpin, "Three Models of the Future," International Organization (Winter 1975).

5. Krasner, Defending the National Interest, p. 10.

6. Ibid., p. 11. White House will be used to refer to those units of the Executive Office of the President that might play a role in the foreign policy process. These would include, in addition to the president and any personal representatives he may choose to act in his behalf:
   The National Security Council
   The Office of Management and Budget
   The Council on Economic Policy
   The Council of Economic Advisers
   The Office of Special Representative for Trade Negotiations
   The White House Staff

   See Marion Irish and Elke Frank, U. S. Foreign Policy (New York: Harcourt Brace Javanovich, 1975), p. 200. The role of the Congress is much less predictable, and may at times emerge to support objectives similar to those of "the state." At other times, members of the Congress may act to support more particularistic interests. In any event, the role of the Congress has remained relatively limited, and is most effective in its ability to prevent certain foreign policy activities, rather than initiate them.

7. Ibid., p. 12.

8. Ibid., p. 13.

9. Ibid., pp. 13-14. For major statements of Morgenthau's theory, see, for example Politics Among Nations (5th edition, New York: Alfred A. Knopf, 1978).

10. Morton H. Halperin, Bureaucratic Politics and Foreign Policy (Washington, D. C.: The Brookings Institution, 1974), p. 11.

11. Krasner, *Defending the National Interest*, p. 15.

12. *Ibid*., p. 16.

13. Franz Schurmann, *The Logic of World Power* (New York: Pantheon Books, 1974), pp. 39-40.

14. David Gold, Clarence Lo, and Erik Wright, "Recent Developments in Marxist Theories of the Capitalist State," *Monthly Review* 27 (October 1975), p. 34.

15. *Ibid*., p. 36.

16. Krasner, *Defending the National Interest*, pp. 21-22, 25.

17. Gold, Lo, and Wright, "Recent Developments", pp. 31-32.

18. Krasner, *Defending the National Interest*, p. 26.

19. Richard J. Barnet and Ronald E. Muller, *Global Reach* (New York: Simon and Schuster, 1974), p. 75.

20. Arthur F. Bentley, *The Process of Government* (Chicago: University of Chicago Press, 1908). For bureaucratic politics approach see, for example, Graham T. Allison, *Essence of Decision: Explaining the Cuban Missile Crisis* (Boston: Little Brown, 1971); and Morton H. Halperin, *Bureaucratic Politics and Foreign Policy* (Washington, D. C.: The Brookings Institution, 1974).

21. Halperin, *Bureaucratic Politics*, p. 4.

22. *Ibid*., pp. 4-5.

23. J. David Singer, "The Level-of-Analysis Problem in International Relations," in James N. Rosenau (ed.), *International Politics and Foreign Policy* (New York: Free Press, 1969), p. 22.

24. John Spanier and Eric M. Uslaner, *How American Foreign Policy is Made* (New York: Praeger, 1974), p. 21.

25. Richard E. Neustadt, *Presidential Power* (New York: Mentor, 1964), pp. 16, 45.

26. Theodore J. Lowi, *The End of Liberalism: The Second Republic of the United States* (2nd edition; New York: W. W. Norton & Co., 1979), p. 36.

27. Krasner, *Defending the National Interest*, p. 28.

28. *Ibid*., p. 35.

29. *Ibid*., p. 42.

30. *Ibid*., pp. 62-63.

31. Ibid., pp. 70-71.

32. S. Cohen, United States Policy, p. 4. Cohen prefers the term "international" economic policy rather than "foreign" economic policy.

33. James N. Rosenau, The Scientific Study of Foreign Policy (New York: Free Press, 1971), pp. 239-240.

34. Halperin, Bureaucratic Politics, p. 17.

35. See especially D. Gale Johnson, World Food Problems and Prospects (Washington, D. C.: American Enterprise Institute, June 1975); Raymond F. Hopkins and Donald J. Puchala, "Perspectives on the International Relations of Food," International Organization, Fall 1978; and U. S. Congress, House Committee on International Relations, "Use of U. S. Food Resources for Diplomatic Purposes - An Examination of the Issues" (Washington, D. C.: U. S. Government Printing Office, 1977).

36. I. M. Destler, "United States Food Policy 1972-1976: Reconciling Domestic and International Objectives," International Organization, Fall 1978, p. 618.

37. Susan George, How the Other Half Dies: The Real Reasons for World Hunger (Montclair, New Jersey: Allanheld, Osmun & Co., 1977), p. 5. George's book originated as part of a counter-report prepared by the Transational Institute for the 1972 Rome World Food Conference. It remains one of the premier works in documenting the relationship between agribusiness and world hunger.

38. Figures on the percentage of United States agricultural exports controlled by the major grain trading corporations vary. The North American Congress on Latin America (NACLA) in 1975 cited a figure of 85 percent, which included Cook industries. Cook has more recently been replaced in the group by Garnac. See NACLA, " U. S. Grain Arsenal," Latin America and Empire Report 9 (October 1975. More recently, William Robbins cited a figure of about 50 percent. See "Cargill: Big Grain Risk Taker," The New York Times, January 9, 1980, p. D17.

39. NACLA, "U. S. Grain Arsenal," p. 18.

40. Ibid., p. 19.

41. Susan DeMarco and Susan Sechler, The Fields Have Turned Brown (Washington, D. C.: The Agribusiness Accountability Project, 1975), p. 5.

42. Emma Rothschild, "Food Politics," Foreign Affairs 54 (January 1976), p. 290.

CHAPTER 2

FOREIGN ECONOMIC POLICY

The Distinction Between Politics and Economics

Any approach to understanding the use of economic resources by the state for "political" purposes must begin with an analysis of the interaction of politics and economics. The development of a framework for the analysis of United States foreign economic policy is a complex undertaking, as it must account for the interrelationship of four basic spheres - the domestic and international aspects of economic policy, and the domestic and international aspects of foreign policy.[1] It is widely acknowledged that it is insufficient to treat foreign economic policy simply as the economic aspect of foreign policy, as political scientists have tended to do; or as the international aspect of domestic economic policy, as economists have been prone to do.[2]

Although the linkage between politics and economics in the international system is by no means a new idea, it is generally conceded that conceptual frameworks for understanding international political economy have not been given sufficient attention in the twentieth century. Political scientists in particular point to increasing "interdependence" as a major alteration in international politics, and consequently deride the bifurcation of politics and economics.

One source of this "divorce" in international relations theory lies in the nineteenth century liberal heritage of modern political science. A basic tenet of liberal theorists was the separation of the political system from the economic system. According to liberal thought, economic processes operate according to natural laws. A harmonious economic system is one that is best left alone to function according to the natural economic order. World peace and prosperity could be best accomplished by an economic order based on a system of free trade and non-interference by governments. The laissez-faire doctrine of nineteenth century liberalism directly challenged eighteenth and nineteenth century mercantilist theory which viewed the acquisition of national wealth as a basic function of the state. As E. H. Carr stated in <u>The Twenty Years' Crisis</u>:

> The theory of the nineteenth century liberal state presupposed the existence side by side of two separate systems. The political system, which was the sphere of government, was concerned with the maintenance of law and order and the provision of certain essential services, and was thought of mainly as a necessary evil. The economic system, which was the preserve of private enterprise, catered for the material wants and, in doing so, organised the everday lives of the great mass of the citizens.[3]

Although modern Marxist theories of the state are generally believed to provide a thorough analysis of international political economy, E. H. Carr argues that Marx himself contributed to perpetuating the distinction between economics and politics. While he recognizes Marx's monumental contribution in insisting on the prominent role played by economic forces in shaping the political system, Marx continued the liberal distinction between the two systems. According to Carr:

> . . . Marx believed, just as firmly did the <u>laissez-faire</u> liberal, in an economic system with laws of its own working independently of the state, which was its adjunct and its instrument. In writing as if economics and politics were

> separate domains, one subordinate to the other, Marx was dominated by nineteenth-century presuppositions in much the same way as his more recent opponents who are equally sure that 'the primary laws of history are political laws, economic laws are secondary.'[4]

A second source of the dichotomy between politics and economics has been the development of political science and economics as separate academic disciplines, a process which has also been traced back to the nineteenth century.[5] Each developed in isolation from the other, and shared very little in the way of common theoretical foundations. Modern international relations is particularly guilty of de-emphasizing economics, and the work of Hans Morgenthau is in part responsible for this situation. The development of political realism as the dominant paradigm in international politics for at least the two decades following the Second World War resulted in an almost exclusive focus on the political activities of states in world politics.

Based on the concept that international politics is a struggle for power among sovereign states, political realism attempted to isolate for analytical purposes actions that a state performs which are of a "political nature." According to Morgenthau, states also perform functions which are not regarded as political, such as economic activities, which are "normally undertaken without any consideration of power." While a state might pursue a wide range of material objectives in its foreign policy, such goals are achieved by the exercise of political power, defined as the ability to control the minds and actions of other men. Political power enables one state to control certain actions of another state by the exertion of psychological power derived from "the interplay of the expectation of benefits, the fear of disadvantages, and the respect or love for men or institutions."[6]

> The main signpost that helps political realism to find its way through the landscape of international politics is the concept of interest defined in terms of power. This concept provides the link between reason trying to understand international politics and the facts to be understood. It sets politics as an autonomous sphere of action and understanding apart from other spheres, such as economics (understood in terms of interest defined as wealth), ethics, aesthetics, or religion. Without such a concept a theory of politics, international or domestic, would be altogether impossible, for without it we could not distinguish between political and nonpolitical facts, nor could we bring at least a measure of systematic order to the political sphere.[7]

The realist paradigm focuses on the state as the primary actor in world politics, and the striving for power as the means by which all states attempt to achieve their goals. International politics is thus characterized as a struggle for power, which is always the immediate aim of the state. States may attempt to achieve their goals by nonpolitical means, such as economic policies, but the realist paradigm places such activities outside the realm of international "politics." Morgenthau makes a distinction between such policies undertaken for their own sake, and the use of such policies as instruments of political policy. An economic policy undertaken for economic purposes, for example, should be evaluated in economic terms. An economic policy undertaken in order to increase the power of the state, however, should be evaluated in terms of its contribution to national power.

The Subordination of Economics in U. S. Foreign Policy

Two major international developments which occurred around the end of the Second World War acted to further reinforce the separation of international politics and international economics, both analytically and in the thinking of statesmen. First was the construction in the West of an economic order, known as the Bretton Woods system, which established a set of rules and institutions based on broad agreement among the industrialized states on international economic matters. At the same time, the Soviet Union was consolidating its position in Eastern Europe, and established under its political and military hegemony a stable economic system which was largely self-contained and isolated from the West. The countries of the Third World remained economically dependent on the industrialized states as an integral but subordinate sphere of an international economic order which was essentially dictated to them.[8]

Beginning with the signing of the Bretton Woods Agreement in 1944 the Western industrialized states, under American leadership, set out to construct an international economic order based on their shared belief in capitalism and liberalism. While the states of North America and Western Europe maintained somewhat different domestic economic systems, particularly with respect to degree of government involvement in the market, all shared a general faith in private ownership and the market mechanism. Internationally, the industrialized states believed in a liberal system of commerce designed to ensure the expansion of free trade. Their aim was to avoid the intensive economic nationalism which had characterized the previous period, which had resulted in a deterioration of world trade and led to a global economic depression. In addition, it was widely believed that the competition fostered by the "begger thy neighbor" foreign economic policies of the 1930s had contributed to the outbreak of World War II. World peace and prosperity, therefore, could be best achieved through mutual economic cooperation.

As embodied in the General Agreement on Tariffs and Trade (GATT), three broad areas of cooperation were specified. First, general rules managing trade relations were established. Second, all parties agreed to lower trade barriers, particularly tariffs and import quotas. Third, a system was established for the adjudication of misunderstandings or disagreements among parties.[9] In essence, the establishment of a world economic order signaled the recognition that maximization of economic welfare could be better achieved through mutual cooperation than by the pursuit of individual domestically oriented preferences. The Bretton Woods system also served, of course, to preserve the position of the wealthy countries of the world from any competition they might encounter from outside.

The international economic order established under the Bretton Woods system was highly successful in governing international economic relations among the major non-Communist states. National trade barriers were lowered, and the industrialized economies prospered at an unprecedented rate. In accordance with these developments, conflict over economic interactions between states was minimized, and international economic matters came to be treated by central decision-makers as essentially "nonpolitical." This created what Richard Cooper calls a "two-track" system of international relations, under which trade and monetary issues were regarded as "low politics," meaning that such matters were relegated to lower levels of decision-making authority. By minimizing the interaction of political and economic issues, the Bretton Woods system managed to confine those international economic issues which had precipitated serious political disagreements in other periods - access to markets, access to materials, and other nationalist protectionist policies - to the economic arena, "in short, to keep trade on its own track."[10]

The second development which contributed to the subordination of international economics was the rapidly increasing focus of United States foreign policy on security matters as a result of the emergence of the Cold War in the late 1940s. The American position in international affairs had changed radically from a long isolationist tradition to the undertaking of enormous global responsibilities as the chief leader and defender of the free world. In an extremely short period of time, American central decision-makers were confronted with an array of political and military issues unprecedented in United States diplomatic history: Soviet domination of Eastern Europe; the conflict over the future of Germany; the formation of NATO; the development of Soviet nuclear capability; the victory of the Communists in China; and the Korean conflict. As a result of these circumstances, the attention of officials as well as analysts of international relations came to center on the security related issues of "high politics," further relegating international economic matters to the level of low politics.

There is some disagreement among analysts over the primary objectives of United States foreign economic policy during this period, which is reflected in the two major interpretations of the origins of the Cold War. Revisionist analyses contend that the thrust of U. S. postwar economic policy was directed toward the creation of an international environment conducive to capitalist expansionism. Other observers support the idea that economic policy was primarily intended to support the American resolve to defend the free world from Communist aggression.

The Bretton Woods system of international economic management was dominated from the time of its inception by the industrialized states of North America and Western Europe. The foundation of the system was shared adherence by those countries to capitalism and a liberal international economic order. There is little doubt that the economic system constructed under the Bretton Woods agreement was intended to provide a favorable environment for the Western capitalist states. The United States in particular was in a highly advantageous position to benefit from an expanding global economy. While the economies of almost all the major powers had been severly devastated by the war, the United States emerged with the most powerful economy in the world. As the primary source of international capital and export goods, an international economic order based on the free flow of goods and capital provided the ideal environment for American domestic and foreign economic expansion. As one observer has noted:

> The economic and political preeminence of the United States during the 1940s and 1950s assured the creation of an international economic order that was tailor-made to American interests, however sensible such an economic order might also be, from the perspective of liberal thought, for maximizing world trade flows, global prosperity, and peace.[11]

American interest in the construction of an international economic order favorable to U. S. economic expansion was joined by a growing U. S. commitment to strengthen the free world and Western Europe in particular, in order to fend off the threat of Soviet Communist domination. The primary thrust of American involvement, embodied in both the Truman Doctrine and the Marshall Plan, was a resolve to eliminate the conditions of economic instability which in the past had contributed to the spread of totalitarian regimes. Thus, for at least a short time, policy-makers apparently believed that the defense of the free world could be accomplished through American economic power. The Truman Administration's rationale for the rebuilding of Western Europe reflected American objectives. As summarized by Walter LaFeber:

> A rejuvenated Europe could offer many advantages to the United States: eradicate the threat of continued nationalization and spreading socialism by releasing and stimulating the investment of private capital, maintain demand for American exports, encourage Europeans to produce strategic goods which the United States could buy and stockpile, preserve European and American control over Middle Eastern oil supplies from militant holdings, and free Europeans from economic problems so they could help the United States militarily.[12]

The belief that American economic predominance could serve as a surrogate for strategic commitments was short-lived. The containment of Soviet power supplanted United States international economic objectives as the Cold War began to take shape. LaFeber continues:

> The Marshall Plan now appears not the beginning but the end of an era. It marked the last phase in the Administration's use of economic tactics as the primary means of tying together the Western world to stop Communist thrusts. The Plan's approach . . . soon evolved into military alliances.[13]

The emerging strategic emphasis was soon concretized by a major reorganization of the United States foreign policy-making machinery. In July 1947 the administration passed through Congress the National Security Act which integrated the three armed services into a single Department of Defense under the direction of James Forrestal, the leading proponent of military solutions to the problems of the Cold War. The Act also provided for the establishment of a Central Intelligence Agency, the Joint Chiefs of Staff, and the National Security Council.

During the eighteen month period following the end of the war, United States policy-makers undertook a dramatic reordering of American priorities in the world. As the strategic concerns of the Cold War became increasingly prominent, the United States abandoned any lingering hope of returning to isolationism and assumed the position of defender of the free world against the threat of Communist domination. American commitment to the maintenance of an international economic order designed to preserve world peace was thereafter overshadowed by the primacy of security issues. Thus the Cold War period had a profound effect on United States central decision-makers' perspectives by creating a perceptual screen through which foreign policy matters were consistently interpreted as threats to the national security. National security became the slogan which political leaders used to justify a wide range of actions designed to achieve a variety of objectives. As described by Robert Keohane and Joseph Nye:

> National security symbolism was largely a product of the Cold War and the severe threat Americans then felt. Its persuasiveness was increased by realist analysis, which insisted that national security is the primary national goal and that in international politics security threats are permanent. National security symbolism, and the realist mode of analysis that supported it, not only epitomized a certain way of reacting to events, but helped to codify a perspective in which some changes, particularly those toward radical regimes in Third World countries, seemed inimical to national security, while fundamental changes in the economic relations among advanced industrialized countries seemed insignificant.[14]

The 1970's: Growing Interdependence

Numerous analysts refer to "interdependence" as the salient feature of international politics during the current period. At the beginning of the 1970s, international economic issues appeared to be increasingly intruding into the realm of high politics. Although the era of interdependence was widely greeted as the dawn of a new period in global affairs, a number of important developments, some of which had been evolving for some time, and some of which surfaced in the early 1970s, have served to shatter the illusion of the separation of economics from international politics. The world appears to have entered a new era of international political and economic reality.

The two most frequently cited sources of this alteration in "traditional" international political relations are the diminution of the use of force by the major powers in the nuclear age and the accompanying partial detente; and the rise in the economic content of relations between states.

For some time now most observers have subscribed to the idea that in the nuclear age military force has become less practical as a policy instrument in the relations between powerful states because nuclear weapons are less "usable" than conventional weapons. Similarly, the use of conventional weapons has become less available to states with nuclear capability because of the possibilities of escalation. This is not to discount the psychological function of nuclear force, which of course accounts for the enormous expenditure of resources by the major powers on preparations for nuclear war. Nevertheless, the nature of nuclear power makes it unavailable as a practical instrument of a state's foreign policy. While the mere possession of nuclear weapons undeniably affects the perceived power position of a state, this capability cannot be utilized instrumentally for purposes of achieving limited objectives. As Hans Morgenthau has stated:

> The relationship that existed from the beginning of history to 1945 between force as a means and the ends of foreign policy does not apply to nuclear weapons. The destructiveness of nuclear weapons is so enormous that it overwhelms all possible objectives of a rational foreign policy. If they were used as instruments of national policy, nuclear weapons would destroy the tangible objective of the policy and the belligerents as well. In consequence, they are not susceptible to rational use as instruments of national policy.[15]

Another factor which has acted to limit the importance of national security issues is the diminished perception of fear of attack among the capitalist states as East-West tensions have abated. Although the maintenance of security remains a major function of United States foreign policy, the posture of confrontation between the superpowers appears to have diminished considerably since the Cuban Missile Crisis experience. Particularly since the 1970s, a degree of limited detente seems to have replaced the tension which characterized the less stable Cold War period, resulting in what Zbigniew Brzezinski refers to as the "partial codification" of the competitive relationship between the United States and the Soviet Union.[16]

As a result of this perceived decline in the security dimension of international politics, other state objectives have risen to the top of the United States foreign policy agenda. Beginning in the early 1970s, international economic issues in particular appeared to be increasingly "intruding" into the realm of high foreign policy. Bergsten, Keohane and Nye support this contention:

> Survival is the primary goal of all states, and in the most adverse situations, force is ultimately necessary to guarantee survival. Thus military force is always a central component of national power.
>
> But insofar as the perceived margin of safety for states widens, other goals - such as economic welfare, political autonomy, and status - become relatively more important. This is in fact the situation at present, as widespread perceptions of detente have rendered quite low most countries' fear of any use by the major military powers.[17]

In addition to this decline in military security concerns by the United States and the other capitalist countries, non-security interests and especially international economics have become significantly more politicized as a result of their increased importance in their own right. By the early 1970's it became clear that a transformation was occurring in the international economic order perhaps equal to the historic alterations which took place following the end of the Second World War. The most significant of these developments included the decline of the U. S. balance of payments position, which in 1971 registered the first deficit of the twentieth century; the emergence of Japan and Western Europe as major competitors for economic leadership of the Western world; the U. S. decision to devalue the dollar, which effectively signaled the end of the Bretton Woods system of international economic order; the transformation of the United States into a resource-dependent state; the emergence of demands by the Third World for a radical restructuring of the international economy; and the appearance of serious conflict among the United States, Europe and Japan over economic issues, in particular national trade policies.

American political and military predominance was a key factor in providing the basis for the establishment of the Bretton Woods system of international economic order beginning in 1944, and that system in turn became essential to the maintenance of such predominance in the following decades. The U. S. position of economic and political power at the end of World War II prompted American decision-makers to assume the responsibility of leadership of the Western world, an action which they had failed to take after World War I. Western Europe and Japan had been severely devastated by the War, and turned toward American leadership to help them rebuild their economic strength and provide international stability. Thus, United States power - politically, militarily, and economically - provided the cornerstone of the Western political and economic order which dominated the international system for twenty-five years.

By the 1970s, however, most observers agreed that the Bretton Woods system of international economic management had collapsed.
American objectives throughout this period were based on a complementary set of favorable political and economic conditions in the international system. Both policy areas were utilized instrumentally in order to achieve objectives in the other area. Thus the continually deteriorating U. S. balance of payments position and a weakening dollar throughout this period were viewed as the price which had to be paid in order to maintain a favorable international political order. By the mid-1960s, however, both Western Europe and Japan had experienced a dramatic economic resurgence, a development which served to seriously undermine the economic basis of American leadership of the Western world. At the same time, a relaxation in East-West tensions, which had provided the political and military basis for U. S. leadership since the end of World War II, acted to further undermine the rationale for Western cooperation under American management.

By the early 1970s the Western world was experiencing an economic crisis of serious proportions. Massive inflation in Europe, instability of the American dollar, and the results of the Arab oil embargo and subsequent dramatic increases in the price of oil following 1974 proved more than the Bretton Woods system was capable of handling. Thus, the disintegration of the postwar international economic order combined with the diminution of security concerns among the major capitalist states to create a situation in which economic issues have once again emerged as the major concern of the relations between states.

Politics and Economics as Concepts

Politics and economics are ordinarily used as terms intended to designate particular systems and the processes which occur within those systems on both the domestic and international levels. Domestic economics encompasses production, consumption and the distribution of goods and services among individuals or firms which take place within a framework of institutions, including all levels of local, state and national government. International economics includes all aspects of domestic economic activity which are connected in any way with actors and institutions beyond the boundaries of the state, as well as activities between states themselves and between states and international organizations.

Politics is used to designate all those activities between nongovernmental actors and institutions and governments, government to government relations; and on the international level, relations between state governments, and between state governments and non-state international actors, which involve influence and control as both means and goals.

Politics and economics thus involve activities within and interactions among four conceptual areas, or analytical systems: domestic politics; domestic economics; international politics; and international economics.

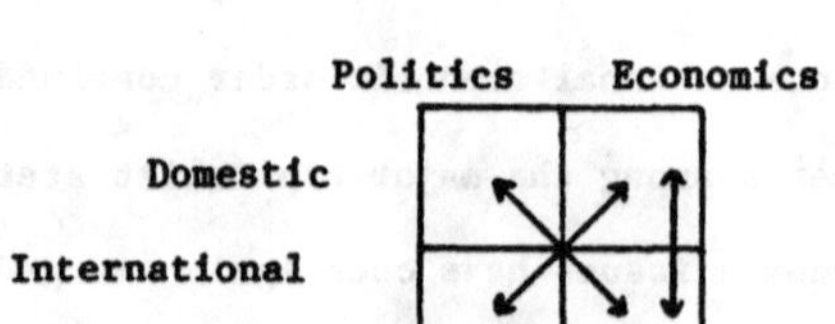

Source: C. Kindleberger, Power and Money, p. 16

According to Charles Kindleberger, international relations may be defined as being comprised of four of the six possible sets of relations as indicated in the Latin square. International political economy is primarily concerned with the relations between international relations and international economics, but is also interested in the connections between domestic politics and international economics, between domestic economics and international economics, and between domestic economics and international politics.[18]

In recent years as international affairs have appeared to become increasingly economic in content, analysts have decried the lack of conceptual frameworks for addressing the interaction between international politics and international economics. According to one observer:

> This situation must be remedied if we are to understand contemporary global politics and the transition of the United States' role in the world. A major obstacle to our understanding of these problems, however, is the poverty of conceptual frameworks with which to address systematically the various interrelationships between international political behavior and international economic behavior. It is important to ask why this is the case and what can be done to increase our analytical capabilities for examining these problems.[19]

Analyses which attempt to explain the relations between states without taking into consideration both the economic and political, as well as the interaction between them must be regarded as being incomplete. Economic activity dominates international relations, as it has for all of modern history. And except for the laissez-faire interlude of the nineteenth century, and during periods of major international wars, economics has normally been recognized as a part of politics.[20] This is not to suggest that either as natural orders or as analytical frameworks one dominates the other, but rather that they are more usefully conceived on different conceptual levels.

Economic activities and objectives are a part of international politics, as are military affairs. Military activities are considered within the boundaries of the political system because, at least in recent history, such activities have been managed by the state. Economic activities, on the other hand, since the nineteenth century have been primarily conceived of as a natural order operating according to its own laws. This doctrine dates to the laissez-faire liberal theories on which classical economics is based. Laissez-faire doctrine further believed that the natural economic order functioned most efficiently when governments did not interfere with it. The nineteenth century theoretical basis for the bifurcation of politics and economics received renewed vigor following the political and economic developments after World War II. This was even further reinforced by the dramatic growth of multinational corporations and other non-state international actors in th 1960s, and subsequent predictions concerning the demise of the nation state as the primary actor in international affairs.

The Interface Between Politics and Economics

Arguments which stress political or security motivations and those which point to economic incentives to the exclusion of the other as frameworks for the analysis of state behavior in the contemporary international system cannot adequately explain the actions of states. Political and economic factors cannot be neatly separated either in the foreign policy decision-making process or in the structure of the international system.

"Radical" or leftist perspectives attempt to portray United States foreign policy as being primarily determined by the economic interests of American corporations and their representatives in the decision-making structure. Conservative perspectives, on the other hand, emphasize security motivations and the struggle for power as the major consideration in the formulation of foreign policy. Such distinctions between politics and economics are not intended to portray the motivations of policymakers, who ordinarily possess an array of objectives, and probably perceive interconnections between motives. Neither do such distinctions portend that there exists an isolation between politics and economics at the systemic level.[21] Such questions as: "Was the United States in Vietnam for political or economic reasons?"; or whether an effort to capture a foreign market is motivated by political or economic objectives, are rather pointless on any level. Conceptual frameworks which are based on this distinction between politics and economics, either on the motivational level or on the international system level should not be construed as attempts to depict the real world. Bergsten, Keohane and Nye have argued on the need for an understanding of the political content of international economics:

> Politics and economics are interwoven strands in the fabric of world order. Two world wars, a depression, and the cold war have made us well aware of the important causal effects of each on the other. Unless definitions of <u>politics</u> and <u>economics</u> are arranged so that one category necessarily includes all fundamental phenomena, neither economic nor political determinism can explain events successfully.[22]

The behavior of states in the present international system can be more usefully understood through conceptual frameworks which examine the relationship between international politics and international economic policies, than by approaches which are based on the existence of isolated political and economic spheres. There is a need to redirect analysis toward this emerging reality so as to address the interface between politics and economics. First, on the motivational level, central decision-makers do not ordinarily possesss objectives which can be realistically classified as being either exclusively economic or exclusively political in content. Policymakers normally assume positions based on some combination of personal beliefs, images of the international environment, and an imprecise set of organizational influences. Second, political activities are often delimited by economic capabilities, thus making a state's overall position of power directly dependent on its economic resources. Furthermore, national political objectives are being pursued with increasing frequency by economic instruments.

Third, the international economic order is determined, to a large extent, by the existing international political system. The economic order which emerged following World War II was primarily the product of the political bipolarization between the East and West. The breakdown of the postwar international economy was, more than any other factor, the result of political changes in the world.

> In the post-World War II period, a new political and economic system based on the hostile confrontation of two superpowers emerged. Politically, the new system was bipolar . . .
>
> This international political system shaped the postwar international economic system. For political reasons, the East and West were isolated into two separate economic systems . . . . Thus, politics have shaped economics in the postwar era. And in the 1970s, as before, it is to a great extent the changing political scene which caused the breakdown of the postwar economic system. The decline of American power and increasing pluralism in the West, a superpower detente, and a new political consensus in the underdeveloped countries are now leading to a new international economic system.[23]

There is also, of course, a reciprocal aspect to this relationship. While political concerns provide the major determinant of the structure of the international economy, the economic process, in turn, serves to transform the political system by redistributing wealth, and thus power as well. "Thus," Robert Gilpin states, "the dynamics of international relations in the modern world is largely a function of the reciprocal interaction between economics and politics."[24]

Finally, all international relations, whether their substance is diplomatic, military, or economic, are political relations. International economic transactions of any sort involve achieving interests, and may therefore be regarded as goal-seeking behavior through interactions among actors ranging from cooperation to conflict. The management of this international conflict and cooperation is the subject matter of international relations.[25]

United States Foreign Economic Policy: The Linkage Between Domestic and International Politics

The foreign economic policy of a state involves the interplay of four analytic domains: the domestic aspects of economic policy; the international aspects of economic policy; the domestic aspects of foreign policy; and the international aspects of foreign policy.[26] This study has already examined the problems with conceptual frameworks which attempt to isolate economic and political phenomena. Still another dilemma encountered in any examination of foreign policy is the increasingly tenuous separation between domestic issues and foreign policy issues. Particularly in the current era of "interdependence," domestic politics appears to "intrude" into foreign policy matters more and more, both within countries and between countries.

The apparent blurring of the modern distinction between United States domestic and foreign policy is the result of two interrelated developments during the last decade. First, as the perceived importance of traditional security issues has declined, economic matters have, in effect, replaced security issues on the national agenda. While United States foreign policy has historically been dominated by economic matters on a day-to-day basis, the combined effects of an established international economic order and an overwhelming preoccupation with strategic matters related to the Cold War acted to relegate economic concerns to a secondary position. Even during the period of the Bretton Woods system, however, while the successfully functioning economic order subordinated economic issues between the United States and the industrialized countries to matters of low policy, American foreign policy objectives in most of the rest of the world, i.e. the developing countries, continued to be primarily economic.[27] While it may not be true that politics "stops at the water's edge" even with respect to questions of national security, international economic issues more directly affect domestic interests and will therefore be more directly determined by the domestic political process.

Second, economic matters have become increasingly "politicized."[28] The last two decades have witnessed the most rapid global economic growth in history, and the modern state continues to become increasingly integrated into the global economy in matters of trade, energy, production, technology exchange and monetary matters. Furthermore, the era of more or less continuous prosperity which had begun during World

War II came to a rather abrupt halt. The major source of American economic difficulties appeared, to many observers, to be emanating from abroad: rapidly increasing economic competition from Western Europe and Japan; demands by the Third World for a redistribution of global wealth under a new international economic order; the first U. S. balance of payments deficit of the twentieth century; and, finally, a dramatic, unprecedented increase in the price of oil by OPEC.

The erosion of common interest and the consequent breakdown of cooperation among the Western industrialized states in international economic relations has resulted in increased politicization of international economic issues both within and among these states. In response to growing scarcity and a major threat to the economic prosperity of the United States, the national government has assumed a significantly broader role in managing the country's economy. During the last decade the American government moved to accept responsibility for an increasingly wider array of economic objectives, domestically as well as in its relations with other states. This has been both as a result of, as well as a significant contributing factor to the breakdown in the distinction between domestic and international economic issues.

Furthermore, as the substance of international politics has more and more come to be dominated by economic matters, foreign policy decisions affect larger numbers of domestic actors to a greater degree in a more tangible manner. As a result, the policy-making process for interdependence issues has become far more complex than has normally been the case for security issues. Foreign economic policy decisions are likely to involve a wider spectrum of government organizations, including many agencies such as the Department of Agriculture, whose central focus and constituency has traditionally been primarily domestic.

The Historical Background of U. S. Foreign Economic Policy

Prior to the period immediately following the end of World War II, the United States had little need for a systematic organization for the management of an international economic policy. Aside from American participation in the financial settlement of World War I, the United States had few objectives and little involvement in the international economy. Commercial interests conducted their international activities with relatively little support or interference from the government. At least until the New Deal era, the extent of state participation in international economic relations consisted primarily of Congressional control of U. S. tariffs. Even in this area, American policy was essentially a reflection of the demands of domestic business interests and was, therefore, protectionist and formulated with little interest in international economic problems. One author has characterized U. S. tradition during this period as "a blend of economic isolation and economic nationalism."[29]

Passage of the Reciprocal Trade Agreement Act in 1934 is cited as the beginning of executive branch involvement in foreign economic policy. Under the Trade Agreements Program, Congress transferred trade regulation authority to the executive branch, with the Department of State being designated to conduct negotiations on tariff agreements, and the Commerce and Agriculture Departments being called upon for consultation. The 1934 legislation, while of limited importance in its own right, marked the beginning of significantly increased involvement by the executive branch in the formulation of national objectives in the international economy. Under the guidance of Secretary of State Cordell Hull, the administration apparently envisioned increased American involvement in world trade as a means of augmenting domestic New Deal programs in stimulating a depressed economy.[30] This linkage between domestic economic and foreign policy objectives marked an important alteration in the evolution of U. S. foreign economic policy.

American involvement in the world economy continued to increase throughout the pre-War period and with the outbreak of war, U. S. economic resources were directed toward helping the allies achieve victory. The White House assumed an unprecedented role in managing the new economic policies and programs aimed at both winning the war and, following victory, preserving the peace. Post-War planning envisioned a stable, open international economic order under the Bretton Woods system behind the leadership and support of a newly activist U. S. foreign economic policy.

As early as 1946, however, the Truman Administration began to fear that the rules and institutions of Bretton Woods would not alone ensure that rapid recovery now perceived as necessary to thwart communism and Soviet expansion.[31]

> Then came the advent of the cold war and the perceived need to urgently strengthen the free world and contain the spread of Communist aggression. The traditional isolationist posture of the United States was discarded. This country's course was clearly set: the bountiful productivity of its large, strong, and undamaged economy would be used in part to finance the establishment of a first line of defense in Western Europe and Japan.[32]

From that point on, U. S. international economic policy became explicitly linked with national security as the Cold War emerged as the predominant conceptual framework for American foreign policy. Whether the Cold War was basically an economic war, with U. S. policy-makers using ideology to implement an expansionist foreign economic policy designed to promote American business interests abroad; or whether economic policy was utilized essentially as an instrument of national security policy attempting to defend the free world, is likely to remain an unresolved controversy. In any event, foreign economic policy was in fact integrated into the framework of the Cold War. U. S. decision-makers undoubtedly did perceive Communist ideology as a threat to American postwar objectives in the world, in political as well as economic terms. The United States assumed primary responsibility for the economic health and defense of the free world, and American foreign economic policy became an integral component of this grand design. "International economic policy, in short, served foreign policy."[33]

The organization for the formulation and administration of United States foreign policy dates essentially from this period, when the predominant concern of central decision-makers was the military and security situation in the world. As America emerged as a global power in the years following the end of World War II, the executive foreign policy bureaucracy underwent substantial expansion, particularly in the area of national security. The most significant organizational development during this period was the passage of the National Security Act in 1947. The Act and its subsequent amendments consolidated the various military departments into a single Department of Defense, established the Central Intelligence Agency, and created the National Security Council as the principal agency for the deliberation and coordination of foreign policy. As defined in the 1947 legislation, the Council's intended function was

> to advise the President with respect to the integration of domestic, foreign, and military policies relating to the national

> security so as to enable the military services and the other departments and agencies of the Government to cooperate more effectively in matters involving the national security.[34]

Membership in the National Security Council included the President, the Vice-President, the Secretaries of State and Defense, directors of the C.I.A. and the Office of Emergency Preparedness, the chairman of the Joint Chiefs of Staff, and the national security adviser.[35] The Council would be called into session by the President, who could also invite other heads of agencies to attend at his discretion.

The National Security Council was created as the principal structure for decision-making during a period when American foreign policy was dominated by political and military issues which appeared to directly threaten the physical security of the country. Furthermore, not only was that threat military in nature, but the primary instruments available for response to the threat were also military. The United States foreign policy structure which emerged, and subsequently underwent dramatic expansion in the twenty-five years following the end of the Second World War was designed for this country's role in a world which in large part no longer exists. The United States has, for the last decade, been formulating and administering its foreign relations within a structural framework based on the problems and capabilities of the late 1940s - when American political, military, and economic power were largely unchallenged.

Contemporary U. S. Foreign Economic Policy

Major changes have taken place in the nature of international relations, as well as in the domestic context within which United States foreign policy is formulated. Increasingly, issues between states are less of a military, security, or diplomatic nature, and more of what is now commonly referred to as issues of interdependence. These issues, which are primarily economic in nature, differ from traditional security-related foreign policiy issues not only in terms of national objectives, but in terms of the number and character of bureaucratic actors which have jurisdiction over such matters in the policy process. Foreign policy is no longer the exclusive, nor even primary reserve of the State, Treasury, and Defense departments, but must be shared with Agriculture, Labor, Commerce, and every other major governmental agency, with the possible exception of Housing and Urban Development.[36]

Furthermore, the nature of involvement by even those agencies which have been regular participants in foreign policy for some time has changed. During the late 1940s, 1950s and 1960s a group of new bureaucratic actors emerged with some degree of responsibility for foreign policy. These included, in addition to the agencies established by the National Security Act of 1947, the Economic Cooperation Administration, which was subsequently reorganized as the Agency for International Development, and the Arms Control and Disarmament Agency. Other existing agencies, which had been essentially domestically oriented prior to this period became involved in foreign policy, and many underwent expansion or established separate foreign policy units.

Involvement in foreign policy by these agencies during the Cold War period was essentially instrumental in support of United States strategic and diplomatic objectives in its role as leader of the free world. Following the successful reconstruction of Western Europe, U. S. decision-makers widened their focus and embarked on a massive counter-communist offensive in the Third World. American economic resources occupied a central role in this effort as a wide range of economic assistance programs were initiated, first in support of a growing number of mutual defense treaties, and then as the United States extended development assistance to Latin America, and the emerging states in Africa and Asia. By the 1960s, responsibility for formulating and administering these programs had significantly expanded the number and power of bureaucratic units within the executive branch which now shared participation with traditional actors such as the State Department in the conduct of foreign policy.

In addition, Congress, which for the most of the post World War II period was not a major participant in foreign policy-making despite its possession of a constitutional role, has lately assumed increased authority in this area. One source of this deepening congressional participation in foreign policy-making can be traced to widespread feelings of executive branch incompetence in recent years. These stem in part from general dissatisfaction with the results of presidential policies in Southeast Asia, the startling revelations of the Watergate episode, as well as a number of perceived foreign policiy "failures" in recent years.

More importantly, however, Congress is playing a more significant role in foreign policy because of the changing substance of international relations. As the content of issues becomes increasingly economic, such issues become more and more integrated into the domestic political process, and thus have created a far more significant role for Congress in the foreign policy process. Furthermore, a rapid expansion of congressional staff over the last decade has made this new involvement possible.

Another consequence of the predominance of interdependence issues has been increased activity by the private sector to secure more influence in foreign policy. Particular domestic interests are far more directly affected by foreign economic matters than by security-related issues, and their influence is enhanced as increased involvement by domestically oriented agencies such as the Departments of Agriculture and by Congress further reinforce the linkage between international and domestic politics.

The United States foreign policy-making structures and processes designed to address the dominant political and military issues of the 1950s and 1960s have not been significantly altered so as to be able to respond to these changes. The present foreign policy environment is placing severe strains on American capabilities, caused most particularly by the absence of high level competence to coordinate economic matters with general (political) foreign policy objectives.

The Organizational Problem

Because of this shortcoming, and persistent reluctance or inability of the Department of State to assume leadership in the economic area, decisions in foreign economic policy tend to vacillate between state institutions, such as the White House and State Department, and domestically oriented agencies. State institutions often lack the specific information and expertise required for handling complex economic issues, and are often unable to adequately assess the domestic implications of these matters. Domestically oriented agencies, such as the Department of Agriculture, will generally command better information and knowledge of a specific issue, but may represent the interests of a particular domestic constituency. In addition, such departments do not possess the capability to account for the requirements of general foreign policy. The fact that United States foreign policy is increasingly economic in content has resulted in a mismatch between the substance of policy and the organization for dealing with it. As stated in a report by the Overseas Development Council written in 1977:

> In the 1950s and 1960s, American foreign policy was dominated by political and military issues bearing, or seeming to bear, more-or-less directly upon U. S. security. More recently, a different range of issues which reflect increasing functional interdependencies between nations has grown in importance. These issues are predominantly but not wholly economic . . . .
>
> The shift in relative emphasis in international politics from the traditional issues of politico-military security to the interdependence issues is imposing additional burdens on American policy-making structures that were already suffering severe strains . . . .[37]

A similar conclusion was reached by the 1975 Commission on the Organization of the Government for the Conduct of Foreign Policy:

> The case (studies on U. S. foreign economic policy) . . . dramatize the most formidable challenge to the organizational architect: the fact that the substance of 'foreign economic' policy is increasingly indistinguishable from what has been called 'political' policy and, perhaps even more important, from what has traditionally been perceived as 'domestic' policy. Without exception, the cases confirm that such distinctions are increasingly inconsistent with reality, but that the institutions created for policy formulation and execution remain largely based upon them . . . .[38]

The formulation of United States foreign economic policy can be usefully understood as a process based on the interaction of two broad groups of objectives: one representing the imperatives of domestic economic policy, and the other representing the exigencies of general foreign policy or national security policy. The basis of foreign economic policy is the establishment of a "basic compatability" between these interests, which at times may appear to be in competition with one another.[39] As stated by Stephen Cohen:

> In a sense international economic policy becomes a means to the end of both domestic and foreign political success. The need to pursue both ends creates the situation whereby domestic economic policy objectives and foreign policy priorities forever are meeting at the intersection of international economic policy. The latter is therefore intrinsically concerned with the nucleus of the other two policy areas. Like domestic economic policy, it is responsible for determining the allocation of scarce resources relative to insatiable demands, as well as influencing employment and price patterns. At the same time, this policy is concerned specifically with national security, that is, preservation of the state's physical integrity and political values . . . .[40]

Development of a framework for analysis of foreign economic policy is therefore extremely complex, as it must take into consideration the process by which domestic priorities are formulated as well as the process by which international priorities are formulated. Furthermore, it must also be concerned with the way in which the two sets of competing interests are reconciled in the political process. Approaches which are based on the conceptualization of isolated domestic and foreign policy processes, therefore, are of limited utility. According to Peter Katzenstein:

> A selective focus on either the primacy of foreign policy and the 'internalization' of international effects or on the primacy of domestic politics and the 'externalization' of domestic conditions is mistaken. Such a selective emphasis overlooks the fact that the main purpose of all strategies of foreign economic policy is to make domestic politics compatible with the international political economy.[41]

The Objectives of U. S. Foreign Economic Policy

In any given situation, United States foreign economic policy may appear to be attempting to achieve a particular objective or group of objectives designed to promote either domestic interests or foreign policy interests. At times, the pursuit of priorities in one of these areas may result in detrimental consequences for the other area. At other times, foreign economic policy may be complementary, and benefit both domestic and foreign policy interests. In almost no situation, however, will the implementation of objectives in one area fail to have any impact on the other area.

In the absence of high-level capability to coordinate these conflicting domains, central decision-makers tend to perceive foreign economic policy in an instrumental sense, as a subordinate to either domestic economic or foreign policy objectives. Both of these areas are operated by large complex bureaucracies which are designed to formulate and administer policies within their own spheres. The central state actors who are primarily responsible for formulating U. S. foreign policy goals remain oriented toward political and security matters within their upper levels. Neither the Department of State nor the National Security Council are organized for consideration of economic problems to nearly the same degree as they are geared toward addressing security issues.

The State Department is technically responsible for the overall conduct of United States foreign policy and has historically coordinated both economic and political relations with other countries. State continues to participate in foreign economic policy, primarily through its Bureau of Economic and Business Affairs, and American embassies abroad still provide important economic reporting. Nevertheless, most observers agree that the Department of State has been reluctant to reorient its focus to the present substance of foreign policy, and has failed to develop a sophisticated capability in the area of foreign economic policy. This situation is additionally compounded by the apparent eclipse of the Secretary of State as the President's chief foreign policy adviser in recent years by the national security adviser, as well as other assorted organizational effectiveness problems of the department. The National Security Council has in recent years become the highest level forum in foreign policy matters, but is decidedly oriented toward addressing political and military matters both as a result of its organization - despite the recent additions of the Secretary of the Treasury and the chairman of the Council of Economic Advisers, the N.S.C. is dominated by the security and defense agencies - and the fact that its role at any given time is essentially a reflection of presidential vision in foreign relations.

Specialized agencies such as Treasury, Commerce and Agriculture are the most prominent among an almost endless list of government agencies which participate to some degree in foreign economic policy formulation and administration. The primary orientations of these departments concern specific aspects of domestic policy. Individual agencies are able to wield considerable influence over decisions within their particular jurisdictions as they normally possess superior expertise and resources within specialized substantive areas. Objectives of the domestic bureaus in foreign economic policy matters are determined to a large degree by their domestic missions. Furthermore, the goals of certain agencies which are major participants in specialized sectors of foreign economic policy are significantly related to their responsibilities to particular domestic constituencies. Nowhere has this situation been more prominent in recent years than in the critical role played by the Department of Agriculture in determining United States food policy.[43]

Foreign economic policy can therefore be viewed as existing as a subsidiary phenomenon, shifting back and forth as an instrument for accomplishing either domestic or international objectives. In any given situation, foreign economic policy is likely to become the "target" of policy-makers in each of these areas who seek to utilize it as means to support their respective priorities. Yet it may also be viewed as existing, in a third sense, as a separate, unique policy area.[44]

It is inadequate to treat foreign economic policy merely as the international dimension of domestic economic policy, or as the economic dimension of foreign policy. For while it is both of these it is more than the sum total of these two parts. Foreign economic policy also exists in a distinct mode as the process which is responsible for balancing and coordinating what may appear to be, at least in an organizational or bureaucratic sense, conflicting national priorities. It must be concerned with the objectives of each of the policy areas from which it is composed - the allocation of scarce resources, the optimization of international trade and investment, maximization of the national wealth, the support of U. S. diplomatic and ideological goals in various parts of the world, and preservation of the territorial integrity of the United States.

Above all this, however, foreign economic policy must be concerned with comprehending the scope and elements of an extremely broad spectrum of issues. It must be capable of coordinating and balancing a wide range of interests and priorities so that the resulting policy reflects the maximizing of national objectives. Edward K. Hamilton, author of the Summary Report on U. S. foreign economic policy for the Commission on the Organization of the Government for the Conduct of Foreign Policy stated:

> . . . it is critical to sensible (foreign economic) policy that national purpose be conceived as a plural construct, made up of matched pairs of opposing concerns in which content and balance change constantly with events, beliefs, and dominant personalities. Thus we must expect constant tension between parochial objectives and those based on concern with the system as a whole, between those which promise short-term advantage and those which look to the longer-term, between those perceived as 'foreign' and those perceived as 'domestic'. . .[45]

Foreign economic policy needs to be comprehended as an instrument of national power, in the same way that "national security policy" must be understood. Conventional analyses of both foreign economic policy and national security policy inevitably fail to distinguish clearly between instruments and objectives. United States objectives in various parts of the world cannot be understood in either strictly economic, or strictly military terms. The national objectives of almost any state in one sense include both economic and military objectives, as well as diplomatic and ideological objectives. Ultimately, however, all of these objectives must be understood as instruments which are employed toward the larger goal of enhancing general national power, which includes economic well-being. Both foreign economic policy and national security policy may be viewed, then, as means by which states attempt to enhance, or maximize general national power, and as elements of general national power.

The Instruments of American Foreign Policy

During the Cold War period military force served as the principal means by which the United States maintained its national security.

Military force, or at least the threat of military force was employed as an instrument to achieve what were, in the immediate sense, military, political, ideological, as well as economic objectives. In the long run, however, all of these instruments can be understood in terms of promoting general national power.

> Political scientists for the past three decades have generally emphasized the role of force, particularly organized military force, in international politics. Force dominates other means of power in the sense that _if_ there are no constraints on one's choice of instruments (a hypothetical situation that has only been approximated in the two world wars). . . .If the security dilemma for all states were extremely acute, military force and its supporting components, which, of course, include a large economic dimension, would clearly be the dominant source of power.[46]

In recent years, however, the perception of military threat has receded in large measure. Economic concerns have replaced the threat to national survival due to the increase in the importance of economic issues in their own right, as well as a result of this decline in the perceived threat to state security. As the nature of state objectives in international politics have changed to include and in fact be dominated by economic concerns, military force has become less useful in achieving these non-military objectives. In addition, the availability and usefulness of military force for the achievement of traditional national security goals, particularly by powerful states, has similarly diminished. Graham Allison and Peter Szanton observe:

> Thus, a paradox emerges: much of our military strength is essential without being usable. Powerful U. S. forces are indispensable. They underwrite the safety of the United States and its major allies from attack, and they reduce the degree to which international relations generally can be dominated by the weight of arms. But those fores provide little positive influence for achieving objectives of our own.
>
> One implication of these changes is a need to reassess the concept of national security. The National Security Act of 1947 viewed security in narrowly military terms and organized a

> military response to protect it. Well suited to its time, that perspective is much less useful now. U. S. policy must seek a broader security, attend a wider spectrum of threats, and develop novel means of countering them.[47]

As this transformation in the traditional effectiveness of military power took place a degree of contradiction developed between U. S. foreign policy objectives in the world, which were decreasingly military in nature, and the means available for achieving them. While it is widely acknowledged that areas of interest act to determine the types of instruments which are developed to support these interests, it is also frequently the case that the instruments most available in turn act to influence the kinds of objectives policy-makers may choose to pursue. The existence of a perceived security threat to the United States, and the assumption of the position as defender of the free world during the cold war period provided justification for the development of a massive military capability supported by a policy-making establishment designed to promote American interests in a world defined in security terms. This national security structure provided the basis for the inclusion of a continually widening area of issues included under the umbrella of the national security symbol. This was in turn cited by policy-makers as further justification for the growth of an even larger national security apparatus, resulting in the kind of cycle which dominated the organization and process of American foreign policy for twenty-five years.

The transformation of issues to an increasing economic nature by the end of the 1960s served to underline the inadequacy of both the narrow, strategic conception of national security, as well as the foreign policy-making structure which had been developed in support of this focus on security matters above all other interests. An effective foreign policy in the present era must recognize that national security contains important non-military components, particularly economic, in addition to traditional military aspects. Secretary of State Kissinger acknowledged this concept in 1975:

> A new and unprecedented kind of issue has emerged. The problems of energy, resources, environment, population, the uses of space and the seas now rank with questions of military security, ideology and territorial rivalry which have traditionally made up the diplomatic agenda.[48]

As a result of these changed circumstances, states have increasingly turned toward economic instruments as a means of achieving their goals in relations with other states. In one sense, economic instruments may be utilized by states as a substitute for military force, in situations where the use of force is unavailable or impractical. The practice of economic warfare has long been utilized by states in conflict with one another.

More important for the present era in international politics is the practice of states employing economic instruments to achieve non-military objectives. The military security of the United States will of course remain as a major concern for central decision-makers. Far more often, however, external threats are likely to be of a non-military nature, involving trade issues, resources, and other such matters which have become more vital in recent years as a result of the increasing interdependence between states. Bergsten, Keohane, and Nye state:

> Where transactions are economic, economic instruments are likely to be used first. They can be wielded by the same bureaucracies that deal with the economic transactions, and they often appear more legitimate to other governments than instruments that appear to "escalate" the controversey to the political-military plane.
>
> Yet much of the complexity, as well as much of the interest, in international political-economic relations derives from the fact that linkages between issue areas, both within the broad arena of economic itself and between economics and security concerns, frequently do take place.[49]

According to the analysis by Bergsten, Keohane, and Nye, further linkages between issue areas are likely to occur due to unequal and different sources of power among states engaged in complex relations. Thus, the Arab states utilized their most clearly available source of power, i.e., economic power by virtue of their oil resources, in order to achieve general foreign policy objectives. More recently, the United States turned to explicit application of its food resources in an attempt to influence Soviet behavior in Afghanistan, apparently perceived as a national security threat.

Thus in the present period of interdependence linkages have increased between the military and physical aspects of security and economic matters, both as elements which comprise general national power, and as instruments employed by states in their relations with other states. There is therefore a need to broaden the concept of national security to include the implementation of economic policy, as well as military and strategic policy for purposes of general national power.

Moreover, it must be recognized that the instruments and elements of national power are not always compatible with one another. Measurement of the success or failure of a particular foreign economic policy cannot be measured entirely in terms of monetary success, but must account for the objectives of that policy in terms of other elements of national power. The withholding of agricultural resources or advanced technology by the United States are examples of the employment of economic instruments for political, or strategic purposes. In the short run, at least, such foreign economic policies may operate to the detriment of economic profit and may, therefore, be measured according to their political success. In the long run, however, employment of national economic power may create a more favorable international environment for the United States, and therefore enhance its position on all fronts, including economic prosperity. According to Benjamin Cohen:

> the rational objective of foreign economic policy is to maximize national power in general, of which economic power is only one single element. National power embodies political, military, geographic, and other elements as well, and while it is certainly true that all of these elements are often mutually reinforcing, it is also true that they are not always perfect substitutes. Consequently, the problem for policy-makers is to estimate costs and allocate means to ends, hopefully to achieve over-all policy goals at the least total cost. At times it may seem necessary to sacrifice one element of national power in order to exploit the more attractive possibilities of another.[50]

Thus in any given circumstance, the purpose of foreign economic policy may not necessarily be directed toward the increasing of economic wealth. Rather it may be employed in an immediate sense in order to achieve what are commonly termed political purposes. Such political concerns, moreover, may have domestic as well as international sources. Broad foreign economic policy programs, such as the disposal of surplus U. S. agricultural resources, serve to support a wide range of political and economic purposes both domestically and internationally, and are often determined by domestic political processes.

Understanding U. S. Foreign Economic Policy Through a "State-Interest" Approach

A mercantilist perspective provides a useful basis for the construction of an alternative framework for understanding the interrelation between policical and economic objectives in United States foreign policy.[51] Under mercantilist doctrine as practiced in the seventeenth and eighteenth centuries, politics and economics were virtually indistinguishable. The state maintained firm control over economic affairs for the expressed intention of increasing its power, defined in terms of maintaining and aggrandizing the national wealth.

The political system during the mercantile period was characterized by a roughly equal balance of state power. Since state structures were relatively weak and armies were small during this period of state formation, diplomatic and military instruments were quite limited as means of competing for power. As Joan Spero has written:

> The impact of the political structure on the economic structure of mercantilism was profound. The economic realm became the main arena for political conflict. The pursuit of state power was carried out through the pursuit of national economic power and wealth; the process of competition, limited by political reality, was translated into economic competition. All international economic transactions were regulated for the purpose of state power.[52]

Under mercantilism, economic policy was conceived as an instrument of political power. According to Heckscher, "mercantilism as a system of power is thus primarily a system for forcing economic policy into the service of power as an end in itself."[53] Mercantilist doctrine thus demonstrated that the pursuit of national power embodied a fundamental economic component. Minchinton refers to this instrumental aspect of economic policy in the mercantilist period as being essential to the process of state formation, during which the emerging states sought to utilize all possible means available in the pursuit of achieving positions of strength:

> To do so they needed to foster economic growth, to raise revenue, and to improve their administrative arrangements . . . .Rooted in practice, mercantilist measures were those economic measures which would lead to the creation and maintenance of a strong nation state. 'Mercantilism' can therefore be described as the striving after political power through economic means . . . .[54]

Jacob Viner's interpretation of mercantilist practice in the seventeenth and eighteenth centuries agrees with the view that the acquisition of wealth by the state was fundamental to the pursuit of power, and additionally supports the idea that power was necessary for the securing of wealth. According to Viner's analysis, the mercantilist state assumed control, or sought to control economic organization and activities in the interest of acquiring and maintaining national power and national wealth. Thus, power and wealth may be viewed as both the means and the end of a strong, independent state.

> What then, is the correct interpretation of mercantilist doctrine and practice with respect to the roles of power and plenty as ends of national policy? I believe that practically all mercantilists, whatever the period, country, or status of the particular individual, would have subscribed to all of the following propositions: (1) wealth is an absolutely essential means to power, whether for security or for aggression; (2) power is essential or valuable as a means to the acquisition or retention of wealth; (3) wealth and power are each proper ultimate ends of national policy; (4) there is long-run harmony between these ends, although in particular circumstances it may be necessary for a time to make economic sacrifices in the interest of military security and therefore also of long-run prosperity.[55]

Economic and political activities, both under the management of the state, thus became essentially inseparable under mercantilist thought. The pursuit of wealth and power were regarded as harmonious in the long run, and improvement or reduction in the standing of the national treasury was generally regarded as an accordant increase or decrease in the overall power of the state. Not unlike the perspective of a "zero-sum" equation of power in contemporary international affairs, economic wealth was similarly viewed as finite, and an increase in either wealth or power of one state was perceived as a loss for others.[56]

The mercantilist contention that economic policy may be essentially compatible with and yet subordinate to the political power requirements of the state marks an important distinction with Marxism. According to a Marxist perspective, the foreign economic policy of the United States is determmined on the basis of the institutional requirements of the capitalist system. Mercantilist doctrine does not dispute that the expansionist objectives of individual American corporations and state priorities of power and wealth are often complementary. On the contrary, American foreign economic policy has often been formulated to support the economic interests of corporations including large-scale overseas expansion, and U. S. diplomatic and military activities have in many instances served vital, instrumental roles in corporate success. There is little doubt the American foreign policy has frequently served the needs of corporations overseas.

According to mercantilist doctrine, however, state objectives in foreign policy do not always coincide with corporate interest. Over the long term, central state actors can be expected to act in support of their determination of the national interest. At times, pursuit of this primary imperative means subordination of economic interests to the political power interests of the state, which may be dictated by ideological objectives, domestic political requirements, diplomatic dynamics, or a perceived threat to the national security. According to Robert Gilpin:

> In general . . . corporate interest and the 'national interest,' as the latter has been defined by succeeding American administrations, have coincided . . . .
>
> The available evidence does not, however, add up to the radical thesis that there is a systematic relationship between American policy and (corporations). It is simply not the case that the imperative of corporate growth and expansion explains the foreign policy of the United States. On the contrary, there are many examples when corporate interest and foreign policy have sharply diverged. In such cases the tendency has been for the larger interests of foreign policy to prevail.[57]

Thus mercantilist doctrine is substantially compatible, at the international system level, with a statist approach to foreign policy formulation. Both approaches conceive of the state as an autonomous, organic unit; and both view the interests of the state, or "national interest," as the primary determinant of state behavior. A major difference betwen mercantilist practice in the seventeenth and eighteenth centuries and the process of United States foreign policy formulation today lies in the fact that central state actors do not possess the autonomy to implement state objectives as policy based solely on their conception of the national interest.

Critical to analyzing the formulation of United States foreign policy through a "state-interest" approach is the recognition that the long term goals of central state actors are determined on the basis of the primacy of national political and economic objectives over the interests of groups within society. At times, the priorities of the state in foreign economic policy may be directed toward increasing the overall national wealth, or even supporting the economic interests of private corporations, when such support appears to be consistent with maintaining a favorable international environment for the economic interests of the state. In other circumstances, however, state priorities may dictate sacrificing economic priorities in order to achieve "political" success. In some situations, national economic and political objectives may be compatible, and can therefore be pursued simultaneously. What is essential in understanding foreign policy through a mercantilist perspective is the idea that in the long run, the primary determinant of state objectives is the maximization of national power and wealth.

The second crucial element in the construction of a state-interest approach is the recognition, as stipulated in a statist perspective, that central state actors do not always succeed in implementing their objectives as policy. According to a statist framework, state actors attempt to pursue priorities associated with the national interest but in so doing, they must often confront powerful groups within society. In the area of foreign economic policy, such groups may have a formidable stake in the outcome of a policy decision, and are likely to possess substantial resources and the ability to oppose state objectives should they conflict with their own particularistic interests.

In this sense, then, understanding the formulation of United States foreign economic policy through a state-interest approach is entirely consistent with mercantilist theory. As outlined by Jacob Viner:

> While mercantilist doctrine, moreover, put great stress on the importance of national economic interests, it put equally great stress on the possibility of lack of harmony between the special economic intersts of the individual merchants or particular business groups or economic classes, on the one hand, and the economic interest of the commonwealth as a whole, on the other. Refusal to give weight to <u>particular</u> economic interests, therefore, must never be identified with disregard for the national economic interest as they conceived it, in interpreting the thought of the mercantilists.[58]

Chapter Notes

1. Stephen D. Cohen, The Making of United States International Economic Policy (New York: Praeger, 1977), p. xviii. Cohen's book is the most comprehensive analysis of the formulation of U. S. foreign economic policy available.

2. In support of this idea, see for example, David H. Blake and Robert S. Walters, The Politics of Global Economic Relations (Englewood Cliffs, N. J.: Prentice-Hall, 1976): Joan Spero, The Politics of International Economic Relations (New York: St. Martin's Press, 1977); and Fred C. Bergsten, Robert O. Keohane and Joseph S. Nye, "International Economics and International Politics," International Organization 29 (Winter 1975).

3. Edward Hallett Carr, The Twenty Year's Crisis, 1919-1939, 2nd edition (New York: Harper & Row, 1946), p. 114.

4. Ibid., p. 116.

5. See especially Benjamin Cohen in the preface to Klaus Knorr, Power and Wealth (New York: Basic Books, 1973); and Spero, Economic Relations.

6. Hans J. Morgenthau, Politics Among Nations, 5th edition, revised (New York: Alfred A. Knopf, 1978), pp. 28-30.

7. Ibid., p. 5.

8. Spero, Economic Relations, p. 2.

9. Richard N. Cooper, "Trade Policy is Foreign Policy," Foreign Policy 9 (Winter 1972-73), p. 20.

10. Ibid.

11. Blake and Walters, Global Economic Relations, p. 14.

12. LaFeber, Walter, America, Russia, and the Cold War, 1945-1975, 3rd edition (New York: John Wiley and Sons, 1976), p. 51.

13. Ibid., pp. 52-53.

14. Robert O. Keohane and Joseph S. Nye, Power and Interdependence (Boston: Little, Brown, 1978), p. 14.

15. Hans J. Morgenthau, A New Foreign Policy for the United States (New York: Praeger, 1969), p. 208.

16. Zbigniew Brzezinski, "U. S. Foreign Policy: The Search for Focus," Foreign Affairs 51 (July 1973).

17. Bergsten, Keohane and Nye, "International Economics", p. 7.

18. Charles Kindleberger, Power and Money (New York: Basic Books, 1970), p. 16.

19. Blake and Walters, Global Economic Relations, p. 3.

20. Carr, The Twenty Years Crisis, P. 116.

21. See Ibid., and Bergsten, Keohane and Nye, "International Economics," for similar arguments.

22. Bergsten, Keohane and Nye, "International Economics," p. 4. The authors cite Joyce and Gabriel Kolko, The Limits of Power (New York: Harper & Row, 1972) as an example of a monocausal argument that economic considerations act to determine U. S. foreign policy.

23. Spero, Economic Relations, pp. 7-8.

24. Robert Gilpin, U. S. Power and the Multinational Corporation (New York: Basic Books, 1975), pp. 21-22.

25. Spero, Economic Relations, pp. 9-10.

26. See S. Cohen, United States Policy and Kindleberger, Power and Money for similar analyses.

27. John R. Karlik, "Economic Factors Influencing American Foreign Policy," in Robert A. Bauer (ed.) The Interaction of Economics and Foreign Policy (Charlottesville: University Press of Virginia, 1975), p. 25.

28. Bergsten, Keohane and Nye, "International Economics," p. 6.

29. S. Cohen, United States Policy, p. 6.

30. Ibid., p. 7.

31. James A. Nathan and James K. Oliver, United States Foreign Policy and World Order (Boston: Little, Brown, 1976), p. 82.

32. S. Cohen, United States Policy, p. 8.

33. Ibid.

34. Graham Allison and Peter Szanton, Remaking Foreign Policy: The Organizational Connection (New York: Basic Books, 1976), p. 75.

35. During the Ford Administration, Congress proposed legislation which would have included the Secretary of the Treasury as a member of the NSC, but the bill was vetoed by the President. In 1976, Congress overrode the veto and Treasury secured a position on the Council, "in an explicit recognition of the extent to which foreign and domestic policy-making are becoming intertwined." In addition, the chairman of the Council of Economic Advisers was added to the

NSC by President Carter. See John Spanier and Eric Uslander, How American Foreign Policy is Made, 2nd edition (New York: Praeger, 1978), p. 53.

36. Allison and Szanton, Remaking Foreign Policy, p. x.

37. Robert H. Johnson, "Managing Interdependence: Restructuring the U. S. Government," (Washington, D. C.: Overseas Development Council, 1977), p. 1.

38. Edward K. Hamilton, Introduction to the Summary Report, "Cases on a Decade of U. S. Foreign Economic Policy: 1965-74," in Commission on the Organization of the Government for the Conduct of Foreign Policy, 7 volumes (Washington, D. C.: U. S. Government Printing Office, 1975), 3, p. 7. A presidential commission established in 1975 "to submit findings and recommendations in order to provide a more effective system for the formulation and implementation of the nation's foreign policy." More commonly, and hereafter referred to as the "Murphy Commission," in reference to its chairman, Robert D. Murphy.

39. Robert J. Katzenstein, "Introduction: Domestic and International Forces and Strategies of Foreign Economic Policy," International Organization (Autumn, 1977), p. 588.

40. S. Cohen, United States Policy, p. 4.

41. Katzenstein, "Domestic and International Forces," p. 588.

42. See note no. 35.

43. For discussion of role of Agriculture Department as representative of U. S. agribusiness interests, see Susan DeMarco and Susan Sechler, The Fields Have Turned Brown (Washington, D. C.: The Agribusiness Accountability Project, 1975); NACLA, "U. S. Grain Arsenal," Latin America and Empire Report 9 (October 1975); Frances Moore Lappe and Joseph Collins, Food First: Beyond the Myth of Scarcity (Boston: Houghton Mifflin, 1977). For a detailed analysis of the Department's objectives in the 1972 grain deal with the Soviet Union, see Dan Morgan, Merchants of Grain (New York: Viking Press, 1979).

When Earl Butz became Secretary of Agriculture in 1971 he relinquished seats on the boards of three agribusiness firms: International Minerals and Chemicals, Stokely Van Camp, and Ralston Purina. Assistant Secretary of Agriculture for International Affairs Clarence Palmby, who negotiated the Soviet grain deal was formerly employed by the U. S. Feed Grains Council, a grain industry trade organization which lobbies to promote the expansion of U. S. grain exports. Immediately following the grain deal, Palmby left the administration to become vice-president of Continental Grain, recipient of the largest agreement with the Soviets. A major purpose of this study will be documenting the role played by Agriculture in other food policy decisions.

44. S. Cohen, United States Policy, pp. 15-16.

45. Hamilton, "Murphy Commission," pp. 7-8.

46. Bergsten, Keohane and Nye, "International Economics," p. 7.

47. Allison and Szanton, Remaking Foreign Policy, p. 55.

48. Henry A. Kissinger, "A New National Partnership," Department of State Bulletin 72, (February 17, 1975), p. 199.

49. Bergsten, Keohane and Nye, "International Economics," p. 9.

50. Benjamin Cohen (ed.), American Foreign Economic Policy (New York: Harper & Row, 1968), pp. 13-14.

51. For major interpretations of mercantilist thought see, Philip W. Buck, The Politics of Mercantilism (New York: Henry Holt & Co., 1942); John Fred Bell, A History of Economic Thought, 2nd edition (New York: Ronald Press, 1967); Eli F. Heckscher, Mercantilism, 2 volumes (London: George Allen & Unwin, Ltd., 1935); and Walter E. Minchinton (ed.), Mercantilism: System or Expediency? (New York: Heath & Co., 1969).

52. Spero, Economic Relations, p. 5.

53. Heckscher, Mercantilism, I, p. 17.

54. Minchinton, Mercantilism, p. vii.

55. Jacob Viner, "Power Versus Plenty as Objectives of Foreign Policy in the Seventeenth and Eighteenth Centuries," in The Long View and the Short (Glencoe, Ill,: Free Press, 1958), p. 286.

56. Edward L. Morse, "Crisis Diplomacy, Interdependence, and the Politics of International Economic Relations," in Raymond Tantner and Richard H. Ullman (eds.), Theory and Policy in International Relations (Princeton, N. J.: Princeton University Press, 1972), p. 129.

57. Gilpin, U. S. Power, p. 144.

58. Viner, "Power Versus Plenty," p. 295.

CHAPTER 3

UNITED STATES FOOD POWER

The utilization of United States agricultural resources for general foreign policy objectives dates back many years. The first documented exercise of American food power appears to have occurred in 1812, when Congress authorized $50,000 in emergency food aid to the victims of a Venezuelan earthquake, reportedly in order to lend support to an unsuccessful revolt against Spain.[1] Since that time, agricultural exports in the form of sales or "aid" have often been instrumental in the achievement of United States national and political objectives.

This chapter will examine the "base" of United States food power. First, it will analyze the scope of American agricultural resources, and how the supply of food commodities is determined by factors in both the domestic and international environments. Second, it will provide a description of the means by which central state actors are able to exert control over agricultural resources, on those occasions where they seek to use food as an instrument to accomplish state objectives.

The role of the U. S. government in facilitating the disposal of the almost continual surplus of food produced by American farmers also has an extensive history. During World War I when war activities severely suppressed grain production in Europe, the United States, under the direction of Herbert Hoover, recognized the opening up of a major market and attempted to sell its grain to both sides. When the British objected to providing such food shipments to Germany, Hoover was able to evade Britain's obstacle by creating the purportedly charitable Commission for Relief in Belgium. In fact, Belgium had no food shortage, and was the most productive country in Europe at the time. U. S. food shipments were simply diverted from the point of their arrival to both France and Germany until the time of American entrance into the war.[2]

Following the end of the war, Hoover turned to America's enormous grain reserves in an attempt to achieve what would later become the major themes of U. S. food policy - the combining of disposal of surplus grain as a means of supporting prices; and its use as an instrument in advancing general foreign policy goals, and in particular opposing Communism.

> Hoover was all for starting aid to the vanquished, who still had gold reserves, while America still had eighteen million tons of surplus wheat. He wanted to make a good deal to avoid plummeting prices in the U. S.; he also wanted to fight the 'collectivist infection' - a pre-Cold-War term for the 'Red menace' - which he feared would spread throughout Europe. All these goals were attained: Hoover sold U. S. wheat and summarily forced settlements of several European disputes simply by threatening to cut off food aid to the party of whose politics he disapproved. He was instrumental in the overthrow of Bela Kun's government in Hungary - and immediately afterwards resumed food aid to that country. He 'suggested' that the Poles accept his choice of Paderewski for their Premier, in which case they might expect increased food shipments.[3]

The term "food power" is generally used to refer to the influence and control one state may exercise over another due to its position as an exporter of food resources to that state.[4] References to food power are normally confined to the utilization of agricultural resources for "political" purposes - i.e., diplomatic or strategic objectives sought by one state in its relations with another. A January 1977 study prepared for the House Committee on International Relations by the Library of Congress offered the following definition:

> Food power is the diplomatic influence that a food-exporting country exercises over the decisions and activities of other nations either because of the control (whether actual or perceived) that the exporting country has over a specific market or segment of a market, or as a concomitant to the ability of the food-exporting country to provide food aid to needy nations.[5]

According to the House study, the United States has available to it a number of policy options with regard to any commodity over which the government may assert market control. The very fact that other states perceive some sort of dependence on a given commodity becomes a factor in a wide range of situations in which states interact with one another. With specific regard to agricultural resources, areas of influence might include normal diplomatic exchanges regarding commercial agricultural trade; more favored trade relations as a result of a quid pro quo for support or concessions in other areas; the withholding of food sales or food aid to certain states in an attempt to influence the behavior of that state in a particular situation; to influence the course of events within that state; or in order to indicate displeasure or dissatisfaction with a particular situation over which that state is involved; as well as the granting of food sales or food aid, or even increasing the quantity of sales or aid in order to extract concessions or support from another state.[6]

The Base of United States Food Power

The ability of American agriculture to produce grain in quantities necessary for satisfying domestic demand as well as the production of an enormous surplus on a dramatically regular basis is well established.

Over the last two decades, United States grain exports have accounted for at least thirty-five percent of total world exports, and in some years as much as fifty percent. Thus the American position of dominance in grain exports is not a recent development, but rather has proved to be quite consistent over a long period of time (see Appendix A, Table 1). U. S. grain production as a percentage of world production has similarly remained fairly stable over this time period, on the average accounting for about one-quarter of all grain produced worldwide (see Appendix A, Table 2).

William Schneider, writing for the National Strategy Information Center offers a position which is representative of observers who argue that American dominance in world food production constitutes a natural base of national power:

> The United States, in particular, and North America in general, are the principal sources of agricultural commodities for the world market. The U. S. lead in agriculture is greater than Arab dominance of the petroleum market . . . .
>
> As a consequence of unfolding events affecting both the supply and demand for agricultural commodities on a worldwide basis, the United States has - without planning it - acquired an unparalleled capacity for influencing international economic welfare through manipulation of agricultural exports. Stated simply, the combination of an increasing worldwide demand for the agricultural products the United States produces in abundance, and the absence of significant alternative sources of production, will place the United States in a unique peacetime position. We have, in short, an effective near-monoply of the raw materials of subsistence.[7]

Most analysts believe that the fundamental condition which must exist, or at least be perceived as existing by certain actors in order for the United States to be capable of utilizing its agricultural resources as a base of national power is a situation of world food scarcity. According to this widely held view, the overall position of the United States' ability to exercise political power based on particular food commodities will vary depending upon global food supply conditions, increasing in periods of short supply and high prices, and decreasing in periods of surplus and depressed prices. Joseph Willett and Sharon Webster of the U. S. Department of Agriculture typically illustrate this position:

> The U. S. food and agricultural situation is now intimately related to the world food situation, and one's judgments about the potential or limitations of U. S. 'food power' will depend to a considerable extent upon his judgments about the future of that situation. It makes a difference whether or not one foresees increasing shortages and rising prices of food, or the alternative of more stable relations betwen food and other prices. If the world of the future should see food demands continually outrunning supplies, as envisioned by the neo-Malthusians, then the struggle for supplies might indeed put the U. S. in a more commanding, although unsought role, in rationing out food.[8]

A similar analysis can be found in an August, 1974 U. S. Central Intelligence Agency study titled "Potential Implications of Trends in World Population, Food Production, and Climate." The CIA report attempts to link trends in world population growth and predictions about global weather patterns to the ability of the United States to utilize its agricultural resources as a base of national power. The study's conclusions are based in large part on projections that the world's climate will continue to experience a cooling trend, the effects of which would be to enhance global dependence on the United States as the world's major exporter of basic foodstuffs. Even if such climatic change fails to occur as predicted, however, the CIA report contends that the American position of dominance over the international grain supply and trade "is almost certain to increase." According to the report:

> This enhanced role as supplier of food will provide additional levers of influence, but at the same time will pose difficult choices and possibly new problems for the U. S. . . . .

> The growing dependence of poor food-deficit LDC's on imported grain and the continued desire of affluent people to increase their consumption of animal products promise generally strong markets for U. S. grain exports and considerable benefits to the U. S. balance of payments. Moreover, ability to provide relief food in periods of shortage or famine will enhance U. S. influence in the recipient countries, at least for a time.
>
> This dependence is also likely to lead to resentment of the U. S. role on the part of the dependent countries. Nevertheless, many will find it expedient to accommodate U. S. wishes on a variety of issues.[9]

Moreover, should climatologists who predict a cooling trend in the world's weather prove correct, as the CIA study believes, the position of the United States as the world's pre-eminent food power would be enhanced significantly. The study continues:

> In a cooler and therefore hungrier world, the U. S's near-monopoly position as food exporter would have an enormous, though not easily definable, impact on international relations. It could give the U. S. a measure of power it had never had before--possibly an economic and political dominance greater than that of the immediate post-World War II years . . . . In bad years, when the U. S. could not meet the demand for food of most would-be importers, Washington would acquire virtual life and death power over the fact of multitudes of the needy. Without indulging in blackmail in any sense, the U. S. would gain extraordinary political and economic influence. For not only the poor LDC's but also the major powers would be at least partially dependent on food imports from the U. S.[10]

In any event, one certain consequence of a scarce world food situation is a high monetary price for food in the international market. The price of a commodity, normally determined by supply and demand, may similarly reflect the potential effectiveness of the commodity as a source of political influence. If supply is short and demand high, a commodity is likely to be capable of demanding a high price in monetary terms, and thus may also command a high value politically.[11]

The monetary price factor itself has important implications for both the major food exporting countries and food importing countries, and particularly the food-dependent areas of the Third World. First, as the price of food has risen, agricultural exports have come to constitute an increasingly important source of national income for food exporting states. Thus while food resources have become more important as a source of political power for the United States, its higher monetary value has also served to generate increased national income from exports. This higher financial value for food has in turn acted to place new constraints on its use as a source of national influence, as any interruption in food exports for political purposes may have negative consequences for important domestic commercial interests. In addition to the significance such interests may have in American electoral politics, commercial agricultural interests play an important role in the food policy-making process and may act to oppose any use of food commodities which might jeopardize their economic objectives. These economic constraints on the use of food for political purpose were generally believed to limit the exercise of food power to markets of marginal economic significance. While the logic behind this line of thought still holds, this assumption was severely shaken by the Carter Administration's decision to embargo grain to the Soviet Union in January, 1980.

Second, a scarce world food situation accompanied by high prices is likely to have far more serious consequences for food-dependent countries which experience financial difficulties in satisfying their food requirements. In surplus conditions such as those which characterized the two and a half decades after World War II, poor countries could expect to receive food donations on at least concessional terms for food purchases from the major exporters, as such terms were economically as well as politically advantageous to producers of surplus agricultural resources. Under scarcity conditions and high prices, however, far smaller amounts of food are made available on non-commercial terms. High prices can therefore be expected to shift the effective demand for food exports from food deficit countries which rely on imports to meet the minimal nutritional requirements of their populations, to affluent areas where imports generally serve to enhance already adequate diets. The potential political, economic and social difficulties which a food-dependent country may face if it is unable to obtain adequate food supplies far outweigh the problems created by oil-dependence in the Western industrialized states. While long lines at the filling station are a definite nuisance, there is little comparison with widespread famine and its consequences. Furthermore, the ability of regimes exposed to such food problems to survive may be especially tenuous if these administrations have utilized for political support the provision of "cheap food" to essential sectors of their societies.[12]

The World Food Situation: The Transition from Surplus to Scarcity

Classifying the world food situation at any particular time as one of either "scarcity" or "surplus" is extremely complex, as the international food system involves a wide range of factors regarding the distribution, production and consumption of a variety of agricultural commodities in virtually every corner of the planet. Moreover, each of these factors may vary as a function of natural conditions, or as the result of policy decisions based on an extraordinarily wide range of goals sought by governments. Political foreign policy objectives may be implemented in conditions of either surplus or scarcity, but to differing degrees and among different potential targets. In a surplus world food situation, "food aid" provides the major instrument by which governments may attempt to influence the behavior of others, which are limited essentially to poor, food-dependent states. Under scarce world food conditions, various types of commercial agreements tend to displace most aid, and the range of potential targets might include all states which are to some degree dependent on food imports, no matter what their financial condition.

In the early 1970s, the situation of chronic surplus which had characterized the world grain situation since the early 1950s suddenly changed to a situation of extreme scarcity. Although the severity of the "scarcity crisis" of the 1972-1974 period has subsided, it appears highly unlikely that the world supply situation will in the near future return to the surplus conditions of the 1950s and 1960s. An understanding of the underlying factors behind the food shortages is integral to understanding the base of United States food power. These factors can be divided into three major categories. The first area of alteration which is commonly cited as contributing to the food crisis involves "natural" circumstances involving a decline in the worldwide production of grain. Included here would be unfavorable weather conditions in various parts of the world, particularly the severe drought in Africa; the sharp decline in rice production in Southeast Asia and Korea; the inability of Soviet agriculture to meet its objectives in production; and the failure of the anchovy catch off Peru.

The second area involves "growth" in worldwide demand for grain. It is widely recognized that global dependency on North American grain exports continued to increase dramatically throughout the two decades preceding the world food crisis, and the years 1972-1974 were unusual only in their severity. There exists less agreement, however, with regard to the reasons behind this increased demand. Some observers believe that growth in the global demand for food is largely attributable to rapid population increases in the Third World outstripping the ability of these areas, most particularly Africa and Asia, to feed themselves. Other analysts cite rising affluence in the developed countries, and accompanying demands for higher protein diets (meaning more meat) by the populations of these countries. And still another factor which certain authors cite as contributing to increased demand for food is the success of ambitious market development programs by the United States government and American corporations in various areas of the Third World, most particularly under Public Law 480.

The third area responsible for the dramatic changes in the world food situation in the early 1970s involves major alterations in policy by the food exporting countries, especially by the United States. Included here would be an apparent decision by the Nixon Administration to bring about significant increases in both the volume and price of American food exports. The most important policy changes involved the decision to curtail the various types of food aid, which had earlier comprised the majority of food exports, in favor of commercial sales to paying customers - the 1972 Soviet wheat deal was the most dramatic example; the decision to allow grain reserve stocks to decline to a two-decade low; and the devaluation of the U. S. dollar in 1971 which, although involving other factors, resulted in lowering the price and therefore making U. S. food exports more competitive in the world market. Two years later, however, once this situation of shorter supply and increased or, more accurately, altered demand had taken effect, the price of U. S. grain more than doubled, and by 1974 had increased approximately three times over the 1972 rate (see Appendix A, Table 3).

The "Decline" in World Grain Production

The 1972-73 world "scarcity crisis" has been widely attributed to adverse weather conditions over various growing regions, in particular the Soviet Union, but also including parts of Asia, West Africa, Australia, and Argentina, combined with a decline in the Peruvian anchovy catch.

The fact world grain production declined only slightly in 1972. While the actual decrease in production amounted to only one to three percent less than the previous year, making it the second best harvest ever recorded, the consequences of this decline were slightly more serious than the loss figure might indicate, due to the fact that in previous years production had been increasing fairly steadily at a rate of approximately 2.3 percent.[13] Consumption of this increased production was divided roughly equally among the developed areas, where increased amounts were utilized as animal feed in an effort to increase the meat content of diets; and the developing areas, where increases were consumed by expanding populations. Thus, at most, effective loss on a world-wide basis amounted to approximately five percent, which should not have been terribly significant, as a sizable world grain reserve still existed at that time.

Declines in production over certain areas of Asia and Africa did occur, but these were slight, and could be expected to occur in some parts of the world in normal years. There was a decrease in the size of the Peruvian anchovy catch, which was then utilized primarily as animal food in the United States, but shortages in the catch had been building up for a number of years.

In 1972 the Soviets experienced a grain harvest below the amount they had announced as their anticipated goal, but again, such occurrences were far from unusual in Soviet agriculture. The 1972 Soviet crop amounted to a total of 161 million tons, down 13 million tons from 1971. Although a shortfall of this dimension is not inconsequential, it was less severe than in 1963, when the shortage equalled 30 million tons, as well as 1965, when the shortage equalled 24 million tons. [14]

Thus neither the Soviet shortfall, nor the net decline in food production on a worldwide basis should have had a significant negative impact on the global food supply. In previous years, grain reserves held by the major exporting countries as well as those held by the Soviet Union itself could have been expected to make up a shortage of this dimension. Another alternative, practiced by the Soviets in similar circumstances, would have been to decrease the size of their cattle herds, for which the missing grain was destined. The difference between the 1972 shortage and other bad harvests in the Soviet Union was that the government made a decision to make up the shortfall with imports, and in conjunction with this action, to sharply cut its normal exports to Eastern Europe from about 6 million tons to less than 2 million tons.[15] Thus for the Soviet Union and Eastern Europe combined total imports suddenly rose from roughly 4 million tons in previous average years, to approximately 28 million tons in 1972-73 (see Appendix A, Table 4). This shift was responsible for most of the alteration in the pattern of world grain transfers in that year, and accounted for what was widely perceived as a serious decline in global food production. Severe "scarcity" did occur in certain parts of the world, principally in the food-deficit developing countries of the Third World, which quite suddenly found themselves squeezed out of the market.

It is, therefore, highly misleading to attribute the abrupt changes which occurred in the world food supply situation to natural, weather-related production problems in 1972-73. Although certain areas of the world did experience unfavorable weather and below-average harvests, this situation occurred virtually every year in some areas, and the 1972 harvest was only slightly below the 1971 record. What was unusual, however, was the dramatic alteration in the global grain transfer pattern away from the poor coutries which had come to rely on North American grain for the survival of their populations, to areas which utilized grain for "improving" already adequate diets. Thus, the major changes which suddenly transformed a situation of chronic overproduction and surplus to one of scarcity were almost totally the result of policy decisions taken in the major importing and exporting states.

The "Growth" in Global Demand

Ever since Thomas Malthus issued his famous warning concerning the inevitability that growth in population would someday outstrip the world's food supply, the world food situation has generally been conceived as consisting of the pressure of population increases on the world's food production capacity. The global demand for food has, of course, historically risen along with growth in population. At the beginning of the twentieth century, the world demand for grain increased at the rate of approximately four million tons per year. By 1950, annual growth in demand had risen to about twelve million tons per year. And only two decades later, this annual incease amounted to roughly thirty million tons, reaching a total global demand of approximately 1.2 billion tons by 1971. By 1985, the United Nations Food and Agricultural Organization projects global demand to equal 1.7 billion tons.[16]

While population growth has historically acted to increase the demand for food, a relatively recent development has been the impact of rising affluence in the developed countries of North America, Japan, Western Europe and the Soviet bloc. As each of these areas have attained certain levels of economic prosperity, their populations have come to expect more meat content in their diets. In order to satisfy this demand, substantial quantities of additional grain are required as animal feed far above per capita requirements in areas where grain is consumed directly. One way to appreciate the implications of this phenomena on the global demand for food is by comparing patterns of grain consumption in poor countries with that in affluent countries. In the developing world, per capita grain consumption averages about 400 pounds per year, almost all of it consumed directly as cereals or bread. The average North American, on the other hand, currently consumes about two thousand pounds of grain per year, almost all of it indirectly as meat and dairy products.

Beginning in the late 1960s, growing affluence in the developed countries combined with rapid population growth in the developing areas of the Third World to place unprecedented pressure on the global food supply. Estimates by the U. N. Food and Agricultural Organization calculate increases in world food consumption since this period as being divided roughly equally between the one billion people in the developed world and the three billion people in the developing countries.[18] In a world food situation characterized by large surpluses and low prices, as existed until 1972, this pattern in global increases in consumption probably would have continued. In the situation of scarce or limited supply and high prices which has existed since 1972, however, countries which had become dependent on food donations or concessional arrangements have been unable to compete with wealthier consumers in an almost totally commercial world food market.

Even during the period when the global food problem was overwhelmingly viewed as one of over-production, and concessional programs were utilized by major exporters to dispose of enormous quantities of grain, poor countries never received more than about one-third of total grain exports. At least since the 1950s, the overwhelming proportion of international grain transfers have occurred <u>among</u> the developed countries of the world for the purpose of "improving" diets (see Appendix A, Table 5). Between 1956 and 1960, a period of severe global food surplus, the developing countries of the world received 32.1 percent of world grain exports. Between 1972 and 1976, during a period of global food scarcity, the developing countries received only 30.5 percent of total world exports. During 1972-73, the year of the "scarcity crisis," grain imports by the Soviet Union were equal to almost 90 percent of the imports by all developing countries combined.[19]

The crucial component in understanding the growth in global food demand, particularly since 1972, is the recognition that while population increases in the Third World will continue to generate a physical need for food, this does not necessarily translate as an economic demand for food. Thus the changes which occurred in 1972 were rooted not so much in the "growth" of the global demand for food, as in the alteration in consumption patterns between the rich countries and the poor countries.[20] According to Peter Wallensteen:

> The present scarcity crisis has two basic roots: the growth of affluence in some countries and the growth of population in others. The immediate crisis in 1972 was created by affluent countries, not by demands exerted by the poor countries. As a matter of fact, the world market is governed by monetary purchase power: it is only buyers that possess significant monetary sources that count.[21]

Thus important changes in the patterns of international food consumption were occurring during this period which acted to place new demands on the existing food stocks held by the grain exporting countries as well as on food production. These changes would not have affected the global food supply, however, if the major food-exporting countries, in particular the United States, had not instituted drastic policy changes designed to accommodate this new demand. In addition, it should be noted that these alterations in demand did not occur suddenly, but rather had been taking place over some period of time. The "scarcity crisis" was not the result of some sudden change in the natural environment, but the result of sudden changes in United States foreign economic policy.

A major obstacle to understanding global food scarcity is the persistence of the myth that world supply conditions are determined primarily by the vagaries of nature combined with relentless population growth among the poor people of Africa, Asia and Latin America. In fact, even though thousands of people per day die of starvation or malnutrition-related diseases in virtually every part of the globe, the world's farmers currently produce enough food to supply every person on the planet with more than 3000 calories per day.[22] The problem is of course essentially one of distribution of income, which in turn determines distribution of food.

Any individual who has the financial resources necessary to purchase food can do so no matter what area of the world he or she lives in. Food can and will be made available wherever there is a demand for it. The key to understanding demand, however, is keeping in mind that demand involves not only the desire for something, but the financial ability to pay for it as well (known as "effective demand"). During the 1972-73 world food crisis, those countries which could afford to import food through commercial channels - Japan, Western Europe and the Soviet Union - continued to do so, in some cases sharply increasing their quantities of imports. Poor countries which were unable to compete in the international food market received drastically reduced supplies of food.

World food scarcity appears to be an almost wholly man-made state of affairs. While bad weather is often cited as a major cause of the most recent scarcity crisis, the 1972 world harvest was the second best ever recorded at that time, and only one percent lower than the record harvest of the previous year. Thus the primary factor which determines abundance or scarcity for any area or individual is not the total world food supply. In any area of the world where there exists an effective demand for food, it will exist in sufficient quantities. Alternatively, in those areas of the world where populations are unable to pay for food, "scarcity" will exist. As stated by Frances Moore Lappe:

> The theory that we are now entering the age of inevitable scarcity because our numbers have surpassed some supposed threshold cannot be substantiated. In a world where food stocks are deliberately depleted so that United States grain exports might earn the greatest foreign exchange and where the major headache of hundreds of agricultural specialists around the world is how to reduce mountains of so-called surplus, the notion of scarcity is worse than a distortion.[23]

Policy Decisions by Major Exporting States

The principal factors which act to determine the world food supply situation appear to be policy decisions by those actors who are in positions of being able to control the major elements of the production, distribution and consumption of food.[24] Such decisions are rarely made on the basis of humanitarian need, with the possible exception of providing disaster relief in isolated occurrences. Rather, policy decisions which act to determine the condition of the international food system are more often made on the basis of economic and political objectives of actors who view food much like any other commodity. Thus, situations of world food scarcity, if they can be said to exist at all, result when these actors determine that such a condition better serves their perceived self-interest at any given time. Frances Moore Lappe comments:

> As strange as it may sound, what we are taught to view as scarcity is actually a product of efforts to cope with the problem of overproduction in a world where most hungry people cannot buy the food that is produced.
>
> This crisis of overproduction spawns scarcity-creating solutions: production cutbacks, the planting of nonfood and animal feed instead of food crops, and built-in inefficiencies in the use of what is produced. There is scarcity, but it is not a

> scarcity of food. The scarcity is of people who have either access to the means to grow their own food or the money to buy it.[25]

The single most decisive factor in determining the condition of the global food supply is United States grain export policy. For twenty-five years the primary objectives of American food policy were first, to dispose of surplus grain created by chronic overproduction by U. S. farmers; and second, to develop and expand markets for U. S. food products abroad. Low, stable prices made possible by the regulatory capability of a large grain reserve and the food "aid" program and its various concessional arrangements were instrumental in accomplishing both of these objectives.

In 1972, despite clear evidence of an unusually large impending global demand for grain, the U. S. Department of Agriculture paid American farmers $4 billion to keep 62 million acres out of production. During the summer of that year wheat prices began their sharpest rise in history increasing approximately 300 percent in ten months. In the fall of 1972, in the midst of a presidential election campaign, the Nixon Administration paid farmers to hold an additional 5 million acres out of production.[26]

The food crisis which occurred in 1972 was not only avoidable, but the available evidence suggests that it was in fact planned by U. S. policy-makers. According to Kenneth Schlossberg, Staff Director of the United States Senate Select Committee on Nutrition and Human Needs, the American government could at the very least have interceded to soften the impact of the 1972 world food shortage, but instead followed a course of action which worsened the situation.

> In 1972, the American government was aware of the following facts: the Soviet Union's wheat crop was in trouble from a

winter freeze followed by summer drought; there had been droughts in Australia and Argentina; the monsoon had been light in India; the world fish catch was in a continuing decine since 1969; there had been a dietary revolution in Japan and Western Europe that called for greatly increased grain supplies to meet livestock demands; the Soviet Union had, by 1970, become a net importer of grain under the pressure of a major decision to upgrade dietary standards by increasing meat consumption; world grain reserves were in a steady decline to a dangerous level of only ten percent of annual needs; population was increasing at over 75 million per year; the prospects for famine in the Sahel and Bangladesh were ominous.

The evidence was clear that demands on American grain would meet new heights. A great proportion of these demands would be humanitarian.

All this was known. It was understood. It made no difference to the American policy-making mechanism.[27]

U. S. Food Policy in the 1970s

In the late 1960s, it was becoming increasingly evident that the American position in the international economy was experiencing serious decline. By 1970, U. S. gold reserves had been depleted to half the 1950 level; the currency position of the dollar had declined; and the international balance of payments was shifting steadily into the red. In response to these difficulties, in 1970 the presidential Commission on International Trade and Investment Policy (commonly referred to as the "Williams Commission" after its chairman, Albert L. Williams of IBM) recommended stepping up the two categories of exports where it was believed the United States could earn substantially increased amounts of foreign exchange: advanced technology and agricultural commodities.[28]

The Williams Commission supported the view that a basic source of America's international economic problems lay in its assumption of "primary responsibility for the economic viability and defense of the non-Communist world" at the conclusion of World War II. Despite significant resurgence on the part of Western Europe and Japan in recent years, those countries did not move to shoulder the responsibilities coincident with their improved positions. It appeared, therefore, that it was no longer in America's interest to remain as guarantor of Western interests in the international economy.

An examination of the U. S. balance of payments, the Commission contended, underlined this situation. In 1970, a year when the United States experienced an overall payments deficit of some $3 billion, non-governmental economic transactions actually recorded a surplus of approximately $5 billion. Governmental transactions, therefore, accounted for a deficit of more than $8 billion, almost $5 billion of which was in military expenditures, with the remaining deficit primarily in foreign aid. The report argued, therefore, that the private U. S. economy was "carrying a burden of governmental payments abroad quite disproportionate to any other country in the world."[29]

Basic to the Williams Commission's recommendation for expanding exports of agricultural commodities was the implementation of a free trade doctrine designed to open up protected European and Japanese markets, as well as large scale encroachment into new markets in the Soviet Union and China. In support of such a strategy the Commission cited the principle of "comparative advantage," which argues that the United States and a few other major producers are best suited for producing grain, while the developing countries of the Third World have a natural advantage in growing a few tropical crops. Futhermore, U. S. export agriculture could be best implemented in an atmosphere devoid of trade barriers and farm programs designed to support the incomes of U. S. farmers.[30]

The recommendations outlined by the Williams Commission became an important foundation of the Nixon Administration's New Economic Policy (NEP). Two major steps were taken toward implementing the objectives of increasing both the quantity and price of U. S. agricultural exports, the results of which led directly to the so-called world food scarcity crisis of 1972-73. The first major policy decision was the devaluation of the dollar by eleven percent in August, 1971.[31] With devaluation, American agricultural commodities became significantly more competitive in the international market. Although devaluation initially made U. S. food products cheaper, it also served to greatly stimulate demand in the industrialized countries of Europe and Japan. In the two quarters following devaluation of the dollar, the quantity of American wheat exports tripled and corn exports increased some twenty percent.[32]

The second step taken by the Nixon Administration toward implementing the agricultural objectives of the NEP involved a series of interrelated actions directed toward converting what had previously been a governmentally managed, primarily concessional foreign agricultural market into a more financially prosperous commercial market. In this area, the major actions involved a drastic depletion of the grain reserve stocks; the extension of a sizable food credit to the Soviet Union, thereby creating a massive new export market; and the effective curtailment of concessional food aid to the developing countries of the Third World.

Despite ever-present rhetoric of free trade and private enterprise the American government has been intimately involved in the U. S. farm economy at least since the 1930s. Since that time, government programs have been primarily directed toward the two basic problems of American agriculture - chronic overproduction and the depressed prices which accompany it - which abated during World War II, but returned in the early 1950s, and have continued ever since then.

During the 1950s, the govermentally managed reserve system accumulated enormous stocks of grain, averaging over 100 percent of the world's annual utilization in the period 1954-1962.[33] Originally conceived as a means of supporting farm income by buying off surplus production, these stocks also served as a buffer against price fluctuations, as well as a means of relieving famine in many parts of the world.[34] Despite the success of Public Law 480 as a means of disposing of sizable quantities of grain abroad and attempts to limit production by withdrawing land, the large reserve stocks continued to accumulate, and had the effect of not only stabilizing prices, but depressing them as well.[35]

By the late 1960s, in the face of consistently low prices, the United States, as well as Canada, Australia and Argentina took steps designed to support the price of grain. In addition to minimum price supports which kept U. S. grain prices on par with world prices, both the United States and Canada began to intentionally reduce their reserve stocks by restraining production. In the United States, wheat acreage was reduced from 59 to 48 million acres between 1967 and 1972, and coarse grain acreage was cut from 103 to 96 million acres. Had these reductions not occurred and acreage held at the 1967 level, more than 100 million tons of additional grain would have been produced in 1972, far exceeding the relatively minor shortfall in global production attributed to adverse weather.[36]

While the idling of millions of acres of North American cropland affectively limited further expansion of American and Canadian reserve stocks of grain, the real opportunity to drastically reduce the size of the reserves came in 1972. During the winter of 1971-72 unfavorable weather conditions over the Soviet Union and Eastern Europe had caused a severe shortfall in their winter wheat crop. In July the United States made the decision to extend to the Soviets a $750 million credit to purchase grain over a three year period. Soviet buyers subsequently arranged to purchase an unprecedented 28 million tons of grain in the world market, approximately 20 million of which came from the United States.[37]

The immediate result of the decision to deplete the grain reserve was a dramatic increase in the world price of grain. Average wheat export prices increased from an average of $1.86 per bushel in 1972, to $3.55 per bushel in 1973, and reached over $5 per bushel in 1974 before leveling off.[38] Another important factor since 1972 has been the dramatic fluctuations which now occur in grain prices. Prior to the reduction of the grain reserve, prices rarely varied more than two or three percent between any given years.

The second result of the reduction of the grain reserve and the consequent rise in prices was the abrupt termination of the United States food aid program. Once the burden of carrying huge stocks of surplus grain had been solved, the food aid program lost its primary objective, which was of course disposal of chronic overproduction by U. S. farmers. Between 1965 and 1974, the amount of wheat shipped under the major program of Public Law 480 declined from more than 13 million metric tons, to just over one million metric tons (see Appendix A, Tables 7, 8 and 9). Furthermore, a large portion of what was left of the food aid program by 1974 went to finance the end of the war in Indochina.

Thus the period around 1972 marked an end to the international food system which had existed since the end of the Second World War. United States reserve policy, which had served during these years to keep prices stable, and provide security against famine, changed abruptly. Since 1972 the United States has held only minimal grain reserves and grain prices now fluctuate dramatically. The food aid program, which provided the equivalent of $25-30 billion in food aid during the 1950s and 1960s, now represents only a tiny fraction of American food exports.[39] Agricultural exports now make up the largest single area of foreign exchange earnings for the United States. Frances Moore Lappe concludes:

> What Americans know as the 'food crisis of rising prices,' starting in 1972-1973, was largely the direct and intentional result of United States 'Food Power' policies that hit upon scarcity as a way to increase both the volume and price of agricultural exports . . . .
>
> Food power was a strategy to create demand and raise prices so as to increase the foreign exchange earnings of the United States. The stage had already been set by acreage cutbacks in the late sixties and early seventies to deal with the mounting surplus of grain. The acreage allotment figure for 1970 was only 75 percent of that of 1967; less land was cultivated in 1970 than in 1948-1952. In both 1969 and 1970 the amount of grain that could have been grown, but was not, on land held out of production amounted to over seventy million metric tons - about double all the grain imported annually in the early seventies by the underdeveloped countries.[40]

Other Factors in Considering U. S. Food Power

Particularly since the formation of OPEC and the consequent transformation of petroleum from a mere natural resource into a highly effective political weapon, numerous observers have attempted to analyze the potential of a similar role for North American grain. While it is true that next to petroleum, grain is by value the largest commodity traded in the international market, a comparison between these commodities must take into account important differences between the two as factors in international trade, particularly with respect to their consideration as sources of political power.

At the present time, about fifty percent of all petroleum produced is traded internationally, while the vast majority of grain is consumed within the country in which it is produced. During the 1950s about six to seven percent of all grain produced was traded in the international market, rising to eight to ten percent in the 1960s and about twelve percent in the 1970s.[41] Another important difference between petroleum and grain is the fact that whereas known, retrievable oil reserves are geographically limited, the production of grain can theoretically be initiated or expanded in virtually all areas of the world.

This idea is in fact advanced by a number of observers who believe that the universal capacity to increase grain production acts as a limiting factor on the ability of certain states to withhold grain exports as an instrument of political power. The belief that the present countries now dependent to a significant degree on imported grain could, however, under their present circumstances simply implement a decision to expand their domestic production, whether by increasing the amount of land devoted to producing grain, increasing yield per acre, or by any other similar procedure ignores the social, political, economic and technological circumstances of these societies. (See Appendix B for a discussion of this issue.)

On the basis of data which indicate that only a small percentage of all grain produced is traded in the international market, particularly when compared with petroleum, certain analysts have suggested that the international grain market may be too "thin" to be utilized as a source of influence in a way similar to that now exercised by OPEC. Two significant trends, however, belie this contention. First, important shifts have taken place in recent years in the margins between production and consumption, moving in one direction in the industrialized countries, and the opposite direction in the countries of the Third World. Henry Nau points out that between 1960 and 1976 in the industrialized countries, the margin of grain production over consumption widened from 1.9 to 4.5 percentage points, thus creating a larger surplus available for export. Meanwhile, in the developing countries the margin of consumption over production widened from 1.2 to 2.7 percentage points (see Appendix A, Table 10). In addition, the centrally planned economies, which were roughly self-sufficient at the beginning of this period, shifted to a position where they also consumed more than they produced. The result of these shifts, should they continue, will be that increasingly larger percentages of grain produced will be traded in the international market in coming years.[42]

Second, even though the percentage of grain traded internationally remains small when compared to petroleum despite trends which clearly indicate its increase, this trade is dominated to an extremely high degree by a very small group of countries, composed of the United

States, Canada, Australia and Argentina. Furthermore, two of these countries, the U. S. and Canada, produce about sixty-five percent of all grain exports, with the U. S. alone accounting for more than fifty percent of the world total (see Appendix A, Table 11).

According to Peter Wallensteen this condition of "supply concentration" is, in addition to the primary condition of scarcity, the second requirement for the utilization of any commodity as a means of influence by a state.[43] When the supply of a commodity is concentrated in the hands of a small number of producers or sellers, the creation of a cartel may be possible. If the supply of a commodity is spread out among a number of sources, on the other hand, producers are likely to compete with one another, and any interruption of supply by one state will be picked up by another. In order for a state to exercise food power based on a particular commodity, then, supply must be sufficiently concentrated so that this cannot occur, as obtaining the cooperation of all other producer states with the capability to replace the interrupted supply is highly unlikely.[44]

The factor of "demand dispersion," given the conditions of scarcity and supply concentration acts to further enhance the utility of a particular commodity as a source of political influence. If there exists only a single buyer for a commodity, or perhaps some type of consumer cartel, then that buyer or group of buyers may be able to wield a considerable degree of influence over the conditions of sale of that commodity, including the price. In such a situation, the producer or seller may even become dependent on the buyer. If there are multiple buyers, however, those buyers may have to compete with one another, allowing the producer or seller to determine prices as well as other conditions he may impose as terms of sale.[45]

While the production of grain in quantities sufficient for export continues to be concentrated within this small group of countries, there have been important shifts among the countries which import grain. The states of Western Europe, which were heavily dependent on grain imports in the decade following the end of the Second World War, accounting for approximately two-thirds of world grain imports in 1950, have dramatically decreased their dependence on imported grain. This import market has been more than replaced, however, by a more diverse group of countries, including the Soviet Union, Japan, and more recently China. In addition, most of the countires of the Third World which were recipients of American food aid during the 1950s and 1960s now constitute a major, seemingly unquenchable market for commercial grain exports, the only limitation being their ability to finance their purchases.

The final factor which Wallensteen contends enables the seller or producer to utilize a commodity as a source of political power is the condition which he terms "action independence." This factor refers to the degree of control or supervision the actor is able to exert over the disposal of a particular commodity. It must be of a sufficient degree so that the state is not prevented from utilizing the commodity instrumentally by either internal or external constraints. The nature of these constraints will vary depending upon the commodity, as well as the individual circumstances surrounding use of a single commodity.[46]

With regard to United States food resources, external constraints do not appear to be present, as they are with respect to some other commodities, in the form of external dependencies. Such dependencies might be in the form of imports necessary for the production process, or some dependence on foreign experts or technology, as are often the case with Third World producers.[47]

There do exist, however, two areas of internal constraints which serve to limit American central decision-makers' ability to utilize food resources for political purposes. The first area relates to the reliance on the export of agricultural commodities as a major source of income, both for food producers and sellers and for the overall U. S. balance of payments.[48] Particularly since the transformation of the international food system following 1972, agricultural exports have become the largest single area of foreign exchange earnings for the United States. Furthermore, one of the costs of maintaining a small grain reserve is that under such circumstances, the domestic economy may be more vulnerable to disturbances in the international grain market.

A second, closely related factor which must be considered by central decision-makers in any decision to employ a food embargo involves potential domestic political costs. Domestic opposition to embargoes reportedly became a factor in the 1976 presidential campaign, and were also a consideration in President Ford's rejection of a proposed grain embargo against the Soviet Union which was proposed as a means of influencing Soviet activities in Angola that same year.[49] Positions on the recent grain embargo against Soviet Union for its actions in Afghanistan quickly emerged as an important issue in the 1980 presidential campaign as well.

This same factor may, however, also have the opposite effect, and act to provide the real motivation for a decision to exercise food power in different circumstances. In the case of the January 1980 food embargo directed against the Soviet Union, the need to demonstrate presidential "action" to the American public in the face of Soviet adventurism was undoubtedly an important factor in support of the embargo.

Because of the critical role of food exports in the U. S. balance of payments, it was widely believed until quite recently (this author included) that this factor would make the withholding of food from major markets such as the Soviet Union far too costly. According to this logic, the exercise of food power directed against small or medium sized markets, probably in the Third World, appeared far more likely. Peter Wallensteen supported this view in 1976:

> Agriculture has achieved a crucial position both in the American balance of payments and in trade as a whole. This imposes constraints on the use of exports as a political weapon. A decision to reduce significant amounts of export would have serious repercussions for the American economy. Furthermore it would have direct effects on an influential group of voters and political financiers.
>
> However, both these points are theoretical. They assume that governmental action would affect a major section of the export. This a highly unlikely development. It is more interesting to discuss a situation where the major part of export is directed to 'secure' or 'friendly' markets. Given such a basic security of exports the probability of using food as a weapon towards marginal export markets increases. In such cases, a reduction of small amounts of American exports will have little or no effect on the American economy, while they could be of great significance for small countries with starving populations.[50]

This position might still be a valid one if it were still the case that the American farmer continued to command the same electoral influence as this bloc constituted in previous periods. Although still an element, the number of American farmers has declined significantly and has been largely replaced in terms of both an economic and political factor by corporate agribusiness. The actual motivations for the Carter

Administration's decision to institute a grain embargo against the Soviet Union may never be completely known. It would not appear unrealistic, however, to speculate that in the administration's thinking, domestic political demand for some sort of action against the Soviets simply outweighed other domestic political and economic consequences.

The Means for Implementing Food Power

Unlike most other states, the United States does not maintain direct control over its agricultural trade by means of a governmental agency. Therefore a major area of investigation must be directed toward what methods central decision-makers have available to them should they seek to utilize American agricultural resources for state purposes. Agricultural production in the United States has of course been traditionally held in and controlled by the private sector. The federal government has, however, been an important participant in the management of the farm economy on a major scale at least since the 1930s, and became even more intimately involved following the end of the Second World War.

As stated previously, the roots of government participation in U. S. agriculture can be found primarily in attempts to address the two traditional problems of American farmers - overproduction and depressed prices. Although policies such as price supports and in later years, the withholding of land from production were important ways in which the government controlled American agriculture, it was the grain reserve system which enabled the federal government to exert direct control over the distribution of large quantities of agricultural commodities.

Although the utilization of food aid for U. S. foreign policy purposes dates back over 150 years, such activities first took place on a major scale as a consequence of the federal government's accumulation of massive grain reserve stocks in the early 1950s. Between 1954 and 1972 large quantities of surplus American grain were distributed under the U. S. food aid program which served as a major instrument of American foreign policy.

The decision to drastically reduce the size of American grain stocks, combined with a change in policy whereby those smaller stocks came to be held almost totally in the private sector has resulted in a situation in which the federal government no longer controls meaningful amounts of agricultural commodities. Under present conditions, where the vast majority of food exports take place in the commercial market under the management of private firms, American central decision-makers are considerably more restricted in their ability to utilize U. S. agricultural commodities to support the foreign policy objectives of the state. This section will examine the means avilable to central state actors for utilizing U. S. food resources for state purposes first, through the food aid program, which predominated until 1972; and second, methods by which the government can presently intervene in the commercial market to exert control over food resources.

U. S. Food Aid Programs Prior to Public Law 480

The United States has administered various forms of food aid on a significant scale since the middle of World War II. It has done so through a variety of programs, beginning with the lend-lease program during the war under which more than $6 billion in agricultural products was shipped abroad, mainly in the form of meat and dairy products. Following the end of the war, the United States administered food aid through the Government and Relief in Occupied Areas program (GARIOA), and the United Nations Relief and Rehabilitation Administration (UNRRA).[51]

The GARIOA program was financed by the Department of the Army as a means of helping to meet the costs of U. S. occupation forces and the providing of relief to the civilian populations of the occupied areas. UNRRA was created as an international organization but was in reality an outgrowth of the lend-lease program, with the United States contributing almost three-quarters of the approximately $3 billion in aid distributed by the program. A major portion of this contribution consisted of foodstuffs and animal feed. Due to heightening East-West tensions, UNRRA was terminated in early 1947. In fiscal year 1948, the United States contributed more than $1.5 billion in agricultural commodities through UNRRA, then in its final year; U. S. foreign relief; the International Refugee Organization; GARIOA; Greek-Turkish aid; the interim aid program; and the Marshall Plan, which was just beginning. Also significant during this period was the changing nature of U. S. food contributions. While in fiscal year 1945 only 56 million bushels of wheat and wheat products were shipped abroad by the United States, exports rose to 318 million bushels in fiscal year 1946, 367 million bushels in fiscal year 1947, 479 million bushels in fiscal year 1948, and reached 505 million bushels in fiscal year 1949.[52]

The purposes of the European Recovery Program were first, to assist Europe in recovering from the direct effects of the war; and second, to provide aid so as to enable the European economy to achieve self-sustenance within the projected four-year time span of the program. Particularly during the initial fifteen months of the Marshall Plan, adverse weather combined with a slow rate of recovery of agricultural production to make imports of food a high priority. During this period, 39 percent of the more than $4 billion in aid was agricultural assistance in the form of food, feed, and fertilizers. Over the life of the program, agricultural assistance comprised almost a third of total aid, amounting to the equivalent of some $9 or $10 billion.

Government food aid activities had a tremendous impact on the U. S. agricultural export economy during this period. During fiscal years 1949 and 1950, government funds through the Marshall Plan and GARIOA financed more than 60 percent of total American agricultural exports. And significantly among all food exports, grain had come to occupy an extremely prominent position in government financed food assistance: in fiscal year 1950, 87 percent of all U. S. grain exports are financed through these two programs.[53]

Also during this period, Congress enacted legislation to earmark Marshall Plan funds specifically for food aid. In 1950, the Yugoslav Emergency Relief Assistance Act authorized some $50 million in Marshall Plan funds for famine relief in Yugoslavia during a period when Soviet-Yugoslav relations were cooling off. In 1951, Congress authorized a $190 million loan to India under the India Emergency Food Act enabling India to purchase large quantities of American grain in the face of a serious famine.[54] As a whole, agricultural assistance provided a substantial component to the highly successful American effort in assisting European stability in the immediate postwar period.

The U. S. Congress passed three bills in 1953 which further laid the groundwork for what was to become Public Law 480. In June, Congress enacted legislation which authorized the transfer of one million tons of surplus wheat in order to assist Pakistan, which was reportedly facing a famine. In July, Congress passed a bill authorizing the president to allocate $100 million in surplus agricultural commodities for "worldwide famine relief." That same month, an amendment was added to the Mutual Security Act which allowed U. S. agricultural commodities to be exported to needy countries in exchange for local currencies.[55]

Also in July of 1953, the Agricultural Trade Development Act of 1953 was reported by the Senate Committee on Agriculture and Forestry. Similar to the amendment to the Mutual Security Act, the bill authorized the president to offer for sale surplus American food for foreign currencies. The bill was passed during the following session of Congress as Public Law 480, the Agricultural Trade Development and Assistance Act of 1954. The purpose of PL 480 was initially twofold: First, the bill was designed to expand American agricultural exports so as to reduce surplus food stocks, which depressed farm prices and were costly to store in reserve. Second, the bill was designed to employ American food surplus in order to assist other nations.[56] The program's potential political value was, of course, not overlooked. As Hubert Humphrey observed at the time:

> I have heard . . . that people may become dependent on us for food. I know that was not supposed to be good news, because before people can do anything they have got to eat. And if you are looking for a way to get people to lean on you and be dependent on you, in terms of their cooperation with you, it seems to me that food dependence would be terrific.[57]

Public Law 480

PL 480 initially contained three sections.[58] Title I comprised the part of the program referred to as "concessional sales," under which U. S. agricultural commodities were exported in exchange for the local currencies of the recipient countries, known as "counterpart funds." Such sales were normally financed on terms which were more favorable than those available through regular commercial sources. About seventy percent of all agricultural commodities transferred through PL 480 have occurred under Title I, making it by far the most important part of the food "aid" program. Between 1955 and the end of 1965, the U. S. sold more than 9.3 billion in agricultural products under Title I.

Title II comprised the donation or grant part of the program. Title II authorized grants of surplus food "to friendly nations for famine or disaster relief or directly to needy peoples of countries without regard for the friendliness of their governments."[59] About half of the agricultural commodities made available under Title II have been distributed through international relief agencies, with the remainder channeled through multilateral relief organizations such as the United Nations World Food Program.

Title III, which has been the smallest section of the program, provided for the barter of surplus American agricultural products in exchange for raw materials from other countries, which were intended to be primarily strategic in nature.

Since its creation in 1954, Public Law 480 has been reorganized, amended and extended several times. In 1966 however, due in large part to problems associated with rapidly accumulating stocks of local currencies by the U. S. government, PL 480 was radically restructured and extended as the Food for Peace Act. While the stated purpose of the new act no longer included disposal of surplus agricultural commodities as PL 480 had, market development and expansion of agricultural trade remained primary objectives. Also, for the first time it specifically stated support of unspecified U. S. foreign policy objectives. The act stated:

> The Congress hereby declares it to be the policy of the United States to expand international trade; to develop and expand export markets for U. S. agricultural commodities; to use the abundant agricultural productivity of the United States to combat hunger and malnutrition; and to encourage economic development in the developing countries, with particular emphasis on assistance to those countries that are determined to improve their own agricultural production; <u>and to promote in other ways the foreign policy of the United States</u> (my emphasis).[60]

PL 480 was initially formulated for the expressed purpose of disposing of the enormous surplus food stocks created by dramatic advances in American agricultural productivity which took place during the 1940s. By the early 1950s, this surplus was reportedly costing U. S. taxpayers $1 million per day for storage alone.[61] Under Title I, food-deficit countries were permitted to purchase U. S. food commodities with their own currencies rather than dollars. This was unquestionably the most significant provision of the program, as it was, through this method, able to create a sizable market for American food exports where none had existed before.

During the first five years of PL 480 the United States exported more than $5 billion worth of grain accounting for nearly 30 percent of all food exports for that period. Title I credits, which were provided at only 2-3 percent interest over a thirty to forty-year period provided the major source of the political influence of PL 480. Also, repayments of Title I credits in local currencies became an important source of

U. S. government expenses in foreign countries, without placing a drain on the outflow of dollars.

In addition, under the Cooley Loan Program passed by Congress as an amendment to PL 480 in 1957, 25 percent of these counterpart funds were permitted to be loaned to U. S. corporations in order to help finance their foreign operations. Between 1954 and 1971 when the Cooley Loan Program was terminated, funds were provided to more than four hundred subsidiaries of American corporations to help them begin or expand operations in thirty-one countries.[62]

Still another use of counterpart funds by the American government was the provision of unspecified "economic assistance" to U. S. allies in the Third World, which was often used by recipient governments as a means of supplementing their military budgets. South Korea, behind India the largest recipient of PL 480 funds, reportedly utilized some 85 percent of proceeds from food aid sales for military expenditures. Finally, counterpart funds were used to provide direct military assistance to U. S. client regimes in the form of grants for "common defense." According to NACLA, $1.7 billion had been utilized in this manner by 1971, with two-thirds of that amount going to South Korea and South Vietnam. And although the United States stopped accepting repayment of PL 480 credits in local currencies after 1971, Cambodia and South Vietnam were exempted from this provision.[63]

During the early 1970s, in the face of increasing congressional opposition to the war in Southeast Asia the Nixon Administration turned to PL 480 as an important source of funding for U. S. foreign policy operations in the area. The administration was also experiencing growing difficulty in securing passage of a foreign assistance program which supported increasingly unpopular "friendly" regimes, from Park's South Korea to the Chilean junta. Congress had begun to subject the annual foreign aid legislation, which was of course never very popular, to increasing scrutiny and had moved to place restrictions on the president's utilization of funds. In an effort to obtain the desired funding for its foreign policy objectives, the administration turned to PL 480 as a means of sidestepping congressional interference. According to NACLA:

> PL 480 provided a perfect cloak for U. S. diplomacy. Few people were fully aware of the political dimensions of food aid, and it was in any case difficult to oppose a program ostensibly aimed at getting food to needy people.[64]

PL 480 contained a highly flexible funding mechanism which the administration was able to take advantage of in order to obtain additional funds for pursuing its objectives in Southeast Asia. Unlike other foreign assistance, PL 480 country programs could be funded by submission for congressional approval of only an estimate of a projected annual budget, which could subsequently be revised without additional permission from Congress.

An examination of expenditures under PL 480 in the early 1970s reveals how utilization of the program had changed toward the explicit support of U. S. foreign policy objectives. In 1974, for example, South Vietnam and Cambodia received $499 million in food aid, although the original administration request for those two countries was for only $207 million. Although PL 480 was under the administration of an interagency staff committee, management of the food aid program by this time had been relinquished by the Department of Agriculture, which no longer needed the program for disposal of surplus food.[65] In its place, the Department of State had taken over initiation of requests for PL 480 expenditures, based upon its own foreign policy priorities. The $499 million allocation to South Vietnam and Cambodia comprised more than half of total U. S. food credits for 1974, a year in which the administration's economic aid request for Indochina was reduced more than 20 percent by Congress. Bangladesh, then in the midst of a famine, received $41 million.[66]

Although the Congress has not been generally inclined to take an active role in allocation decisions, it did act to limit the outwardly political component of the food aid program, largely in response to the unprecedented levels of assistance being channeled to Southeast Asia by the Nixon Administration. In 1973, Congress imposed a section on the Foreign Assistance Act which prohibited the use of PL 480 counterpart funds for common defense purposes.[67] The Humphrey Amendment to the 1974 Foreign Assistance Act mandated that no less than 70 percent of food aid go to countries on the United Nations "Most Seriously Affected" (MSA) list, thereby limiting the amount of aid which could be used for strictly strategic-political purposes in non-needy areas to 30 percent.[68]

Accounts of administration efforts to circumvent the 70/30 stipulation further illustrate the overtly strategic nature of the food assistance program during this period. One account reports that the State Department first attempted to have certain non-food commodities such as cotton and tobacco excluded from the restriction.[69] In January, 1975, Secretary of State Kissinger personally appealed to the United Nations to have South Vietnam reclassified as a MSA country so that it would not be covered by the limitation. Having failed at these attempts, the administration then attempted to increase the amount of food aid funds for strategic purposes by proposing that the 30 percent limit apply to total assistance under both Titles I and II. This attempt also failed, rejected by the Senate Foreign Relations Committee. Eventually, the administration circumvented Congress by simply increasing the total food aid budget by $500 million to $1.5 billion (something it had refused to do at the recently convened World Food Conference in Rome). By increasing the budget of all food aid, the administration made available more total assistance for political purposes, while still remaining within the 30 percent guideline.[70]

PL 480 also played a role in U. S. foreign policy strategy in Chile. PL 480 Title I credits were suspended to the Salvador Allende regime, and were then resumed following Allende's removal and the installation of the military regime. In the face of domestic opposition to U. S. efforts to support the Chilean junta, the Nixon Administration utilized PL 480 as a method of sidestepping attempts by Congress to limit economic and military assistance. For fiscal year 1975, Congress cut off all military aid to Chile, and placed a $26 million ceiling on economic aid. That year, the administration increased its original request for Chile of $35 million in PL 480 credits - already the largest amount for Latin America, which was slated to receive only a total of $50 million - to $65.2 million, $61 million of which was under Title I.[71]

Following U. S. withdrawal from Southeast Asia, PL 480 credits were shifted to other areas of U. S. strategic interest. For fiscal year 1975, South Korea's PL 480 credits were doubled from the previous year to $150 million, which accounted for 17 percent of the program total. This increase occurred following efforts by Congress to reduce military aid to South Korea. There is also evidence that PL 480 credits played an important role in Kissinger's Middle East strategy during the same period. Following passage in Congress of the 70/30 stipulation, Egypt was quite suddenly placed on the UN's MSA list and received $120 million in PL 480 credits in fiscal year 1975, making Egypt the third largest recipient of food aid funds that year.[72]

Following subsequent elimination of the MSA criteria Congress attempted to devise other schemes to direct food assistance to the most needy countries. PL 94-161 was passed by Congress in an attempt to establish a per capita GNP assistance factor for food aid. Under the law a 75-25 poor country/wealthy country ratio was established for food assistance. The problems in employing per capita GNP figures as a means of measuring development are widely recognized. It is even less useful as a method of determining food assistance needs, and was never claimed as an indicator of food deficits or surpluses. Neither the MSA scheme nor the per capita GNP plan proved to be effective in attempts to exclude political considerations from humanitarian ones in allocating food assistance. Yet another proposal set before the House of Representatives in 1975 would have linked food assistance to countries with successful population control programs.[73]

From the time of its inception, Public Law 480 experienced severe political problems, and suffered serious shortcomings as a source of United States political power. As stated in the House International Relations Committee report:

> From the beginning, the fact that the primary objective of the program was to export surplus stocks of agricultural commodities acquired through Commodity Credit Corporation price support operations worked against the diplomatic success of the program. The secondary objective, to provide food aid to combat famine and malnutrition was sometimes overlooked.[74]

There were numerous complaints concerning the program. Other food exporting countries charged that PL 480 amounted to subsidized dumping in the Third World. Critics charged that the program was harmful to recipient countries, as it allowed them to avoid any long-term solutions to their food shortage problems by devoting attention to the development of their own agricultural productive capacity. Thus, many recipient countries were able to provide cheap food for their urban populations while avoiding the investment of resources in rural agricultural development necessary for any long-term solutions to their food problems.

Still another problem cited with PL 480, prominent during the early years of the program was the tendency for the U. S. to accumulate large quantities of local currencies. By 1972 for example, India held a local currency debt of more than $3 billion to the United States. This fact, along with considerable resentment of the dependency relationship fostered by the food aid program became a significant source of distress in the relationship between the two countries. In other instances, the nature of the U. S. surplus agricultural commodities was not suited to the needs of the recipient country. Many traditional rice-staple countries, for example, were offered wheat under the program, leading to serious questions regarding the actual beneficiaries of the food assistance. Still other recipient countries felt serious resentment over certain "strings" or preconditions often required for receiving aid under the program, and were viewed as U. S. interference in the internal affairs of these states.

Public Law 480 was a major instrument of government control over U. S. agricultural resources during a particular historical period between 1954, when the program was created as a means of reducing massive food surpluses, and for paving the way for later commercial sales; until the period 1972-1974, when American policy-makers apparently decided that their interests could be better achieved in a commercial food market. During the span of the program, the major objectives of United States foreign agricultural policy - surplus disposal, market development, and support of general foreign policy - were most often complementary. Since the 1972-1974 period, however, with U. S. food policy operating in a tight, commercial market, significant competition has appeared among those interests.

The Commercial Food Market

A major area of investigation with regard to the potential implementation of food power by central decision-makers is how such policies can be administered within a so-called free market economy. As a result of the major changes in United States food policy following 1972, almost all agricultural exports now occur in the commercial market. Only small quantities are transferred through food assistance programs, which for almost two decades served as the major instrument of U. S. food power.

Under present circumstances, therefore, the major private actors involved in the export of U. S. food commodities have come to play a highly significant role both in the formulation of American foreign agricultural policy, and in its implementation. Unlike virtually all other countries which engage in large-scale transfers of agricultural commodities, the United States does not employ a government trading corporation. Rather, a small handful of private grain trading corporations exercise extraordinary monopoly and monopsony control over U. S. food exports.

The grain corporations occupy in a number of ways a rather unique position in both United States and global agriculture. Almost completely unknown prior to the 1972 Soviet grain deal and the events which accompanied it, it is now widely recognized that a handful of these extremely powerful corporations occupy a role in United States foreign agricultural policy comparable or surpassing the "seven sisters" of the oil industry. These corporations - Cargill Inc. of Minneapolis, Continental Grain of New York, Louis Dreyfus & Co. of Paris, the Bunge Corporation of Argentina, and Garnac Grain, an affiliate of Andre Inc. of Switzerland - are instrumental in controlling the flow of grain in every part of the globe. They exercise ownership or control over a vast agricultural infrastructure which begins in the wheat fields of the American midwest and reaches into the foreign markets of Europe, the Soviet Union, China, and most of the Third World.

The grain traders are the major instruments by which U. S. grain is bought from American farmers and exported all over the world. In order that central state actors be able to utilize U. S. agricultural commodities for any sort of national purpose, therefore, the federal government must have some capability to impose controls over the activities of these corporations.

During the period 1947-48, when the United States first began experiencing overproduction of grain on a regular basis, the various food aid programs connected with post-war reconstruction consumed the majority of surplus food, and commercial exports of American food remained small. The primary legislative authority for regulating U. S. exports including agricultural commodities dates back to this period.

The 1949 Export Control Act was the first major economic warfare legislation after World War II. It was designed to be implemented essentially against the Soviet bloc, and gave the president broad authority to limit or cut off completely American exports in order to prevent economic shortages; to support national security objectives; or for general foreign policy purposes.[75]

In 1962 the Export Control Act was amended in order to grant the president authority to control exports in order "to prevent any significant contribution to the military or economic potential of a Communist bloc country through the import of technology from the United States." Under the amended Act, a licensing system was established under the Department of Commerce Office of Export Control, which consisted of two tiers. One level established a general license under which most goods could be exported to most countries without specific application by the exporter. On the other level, specific authorization was required in the form of a validated license from appropriate federal agencies in order to export good to Communist bloc countries.[76]

The new Export Administration Act of 1969 replaced the Export Control Act, maintaining the export control machinery.[77] Although the revised act was apparently intended to expand trade which had no direct military significance, strict controls on exports to such countries as Cuba, North Korea, and (then) North Vietnam could still be enforced under th 1971 Trading with the Enemy act. Other legislation for regulation of trade with Communist bloc countries was retained under authority granted by the Mutual Security Act of 1954, the Agriculture

Control Trade and Assistance Act of 1954, and the Mutual Defense Assistance Control Act of 1951. In addition, the fifteen NATO countries maintain through their Coordinating Committee coordination of controls on trade of military significance between NATO states and the Communist bloc.[78]

Even under circumstances where the Export Administration Act may not be formally invoked, the existence of presidential authority to implement export controls as stipulated in the Act undoubtedly acts to create a source of considerable power for the president. It has been reported, for example, that it was the existence of standby authority as conferred in the Export Administration Act which convinced U. S. based grain corporations to cooperate in the "voluntary" prior approval system established by the Department of Agriculture in 1974. Under this system, which is still in effect, exporters are required to obtain USDA approval prior to concluding any "significant" (50,000 tons or more) grain sales to a foreign country.

When President Carter initiated a grain embargo against the Soviet Union in January, 1980, he did so under the legislative authority conferred by the Export Administration Act. In memorandums to the Secretaries of Commerce and Agrigulture directing them to take measures to implement the embargo, the President stated:

> I hereby direct that you . . . take immediate action under the Export Administration Act to terminate shipments of agricultural commodities and products, including wheat and corn, to the Soviet Union . . . . I am taking this action in the national security and foreign policy interests of the United States. I have determined in accordance with the Export Administration Act that the absence of controls would be detrimental to those interests and that alternative courses of action would not comparably advance them.[79]

The Department of State had additionally confirmed that it regards the 1969 Export Administration Act as the authorizing legislation for the imposition of executive controls on agricultural exports. In response to inquiries regarding the legal authority for the administration's request for the 1975 grain sales moratorium State declared that: "The Department regards the Export Admministration Act as the basic authority for the President to impose domestically enforceable export controls."[80]

Still another source of government control over agricultural exports is the Agriculture Department's Commodity Credit Corporation. The CCC was created in 1933 as part of the U. S. Department of Agriculture, with status as a government corporation with independent spending authority. Under the CCC, a complex of programs were instituted in order to deal with overproduction of agricultural commodities and to support farm income.[81]

In recent years the Commodity Credit Corporation's independent spending authority has been utilized to grant substantial credits to foreign governments, apparently for foreign policy reasons. It was the CCC, for example, which issued $54 million in agricultural credits to the Chilean junta in the period immediately following the coup. It was also the CCC which extended a $750 million credit to the Soviet Union in July, 1972 which enabled the Soviets to finance their massive grain purchases from the United States. CCC credits have also been withdrawn for foreign policy reasons, as appears to be the case in the early 1970s when credits were terminated to the Peruvian regime as part of U. S. policy to slow down economic aid to that country, which had nationalized certain U. S. investments without compensation.[82]

Summary

During the 1950s and 1960s United States foreign agricultural policy consisted primarily of: (a) disposal of surplus commodities and the creation and expansion of overseas markets under food assistance programs; (b) maintenance of farm income and price stability through the government grain reserve system; and (c) support of foreign and national security policy, primarily by furnishing cheap imported food to favored clients. These basic elements normally converged, and little real conflict existed between the major decision-making sectors on foreign agricultural policy.

Since the alterations which occurred in U. S. policy and the international food system in the early 1970s, however, the major objectives of foreign agricultural policy have become considerably less congruent, and significant opposition has arisen among the major interest sectors involved in the formulation and implementation of policy. The direction of food policy now seems to shift back and forth, at times appearing to represent the interests of corporate actors whose goals are primarily economic, and at other times appearing to serve as an instrument of U. S. foreign policy and national security objectives.

This shifting situation in food policy is, of course, a direct function of the nature of foreign economic policy, and foreign agricultural policy in particular. Probably in no other area do domestic and foreign policy considerations interface more directly. As a result, government agencies with highly different missions often compete for jurisdiction over policy. As the Williams Commission concluded:

> The formulation and administration of foreign economic policy is a complex task. It invariably affects other aspects of our foreign relations. National security may be involved; domestic economic policy nearly always is. In consequence, virtually

> every government department and regulatory commission has participated in developing and administering some facets of our foreign economic policy.
>
> This diverse involvement is unavoidable because of the wide variety of interests which must be heard, the specialized knowledge that is required, and the close links between international and domestic economic policies.[83]

The Commission further concluded:

> In no sector of the economy are domestic and international policies more closely related than in agriculture.[84]

These two competing areas of interest are in turn each represented by major sectors of the government which can be expected to act as proponents, or institutional voices for their respective priorities in any foreign agricultural policy decision. (Although individual decision-makers within these sectors may share concerns in terms of either domestic economic or foreign policy priorities, this does not suggest that they will necessarily agree with one another as to specific courses of action.)

It is a basic contention of this study that the direction of United States foreign agricultural policy, i.e., whether political-security objectives or, alternatively, corporate economic objectives underlie policy is determined to a significant degree by the location of a particular policy decision. In certain situations the president, a direct representative of the president, Secretary of State, national security adviser, or other high level official whose priorities relate to general foreign policy or national security matters may either initiate or intervene in foreign agricultural policy. On those occasions where such state actors are able to exercise jurisdiction over policy, economic objectives are likely to be subordinated to overriding political and strategic considerations. Alternatively, when foreign agricultural policy remains within the jurisdiction of the Department of Agriculture, the thrust of policy is likely to be directed toward domestically oriented, essentially economic objectives.

Depending on certain variables, then, the underlying objectives of a particular foreign agricultural policy decision will reflect the priorities of either central state actors, or domestic economic groups, but usually not both. When the process of formulation of policy remains within the jurisdiction of the Department of Agriculture, its direction will be determined by the "normal" organizational processes within that agency and, most importantly, by its organizational essence.

The exception to these circumstances will occur when certain central state actors - the president, the national security adviser, the Secretary of State, or their direct representative - either initiate or intervene in policy and locates, or relocates the decision process within a state agency. As a result, the direction of policy will be directed toward the support of state objectives.

Chapter Notes

1. Susan George, How the Other Half Dies: The Real Reasons for World Hunger (Montclair, New Jersey: Allanheld, Ossmun & Co., 1977), p. 164.

2. Ibid.

3. Ibid., p. 165.

4. The terms "food power" applied to the utilization of American agricultural resources for political purposes is somewhat of a misnomer, as it actually refers only to that area of food commodities in which the U. S. clearly predominates in terms of export production, namely grain. The term "grain" is widely used to include wheat, rice, corn and other cereals; soybeans (which is actually a legume); and products based on grain, chiefly animal feeds.

5. U. S. Congress, House Committee on International Relations, Use of Food Resources for Diplomatic Purposes, (Washington, D. C.: U. S. Government Printing Office, 1977), p. 1.

6. U. S. Congress, Use of Food, p. 5.

7. William Schneider, Food, Foreign Policy, and Raw Materials Cartels (New York: Crane, Russak, and Co., Inc., 1976), pp. 20, 32.

8. William W. Willett and Sharon B. Webster, "'Food Power': Food in International Politics," in Political Aspects of World Food Problems (Kansas State University, July 1978), p. 167.

9. Central Intelligence Agency, "Potential Implications of Trends in World Population, Food Production, and Climate," (C.I.A., August 1974), pp. 34-35.

10. Ibid., pp. 39-41.

11. Peter Wallensteen, "Scarce Goods as Political Weapons: The Case of Food," Journal of Peace Research 13, 1976, p. 278.

12. Cheryl Christensen, "Food and National Security," in Klaus Knorr and Frank N. Tager (eds.), Economic Issues and National Security (Lawrence, Kansas: Regents Press, 1977), pp. 289-290.

13. For analyses of the relationship between world food production and scarcity, see Dan Caldwell, Food Crises and World Politics (Beverly Hills, California: Sage Publishing Co., 1977); George, How the Other Half Dies; D. Gale Johnson, World Food Problems and Prospects (Washington, D. C.: American Enterprise Institute, June 1975); and NACLA, "U. S. Grain Arsenal," Latin America and Empire Report 9 (October 1975).

14. Fred H. Sanderson, "The Great Food Fumble," Science 188, May 1975, p. 504.

15. Ibid.

16. Lester Brown, By Bread Alone (New York: Praeger, 1974), p. 6; James P. Grant, "Food, Fertilizer, and the New Global Politics of Resource Scarcity," Annals of the American Academy of Political and Social Science, July 1975, p. 14.

17. Brown, By Bread Alone, p. 6.

18. Grant, "Food, Fertlizer," p. 14.

19. Cheryl Christensen, "World Hunger: A Structural Approach," International Organization 32, Summer, 1978, p. 758.

20. Raymond F. Hopkins and Donald J. Puchala, "Perspectives on the International Relations of Food," International Organization 30, Summer, 1976.

21. Wallensteen, "Scarce Goods," p. 281.

22. Frances Moore Lappe, Food First: Beyond the Myth of Scarcity (Boston: Houghton Mifflin, 1977), p. 112.

23. Ibid., p. 26.

24. For a structural approach to analysis of the international food system, see Hopkins and Puchala, "Perspectives on Food," p. 597.

25. Lappe, Food First, p. 22.

26. Ibid., p. 322.

27. Kenneth Schlossberg, "United States Food Policy: The View From Congress," In Giulio Pontecorvo (ed.), The Management of Food Policy (New York: Arno Press, 1976), p. 238.

28. Lappe, Food First, pp. 220-221.

29. United States International Economic Policy in an Interdependent World, Report to the President submitted by the Commission on International Trade and Investment Policy (Washington, D. C.: July 1971), p. 7. Cited hereafter as the Williams Report.

30. Ibid., p. 4. For a discussion of the role of the Williams Report in U. S. foreign agricultural policy, see Lappe, Food First, p. 181.

31. The dollar was subsequently devalued an additional six percent in early 1978.

32. NACLA, "U. S. Grain Arsenal," p. 7.

33. Christensen, "Food and National Security," p. 295.

34. Since the end of the Second World War, the world's food reserve capacity has been held in two principal sources: the carryover reserve stocks of grain held by the major exporting countries; and large areas of farmland held out of production as a means of supporting chronically depressed prices. The United States is the major participating country in both these areas. Although many grain importing countries also maintain sizeable grain reserves, these stocks are not normally available for export and therefore do not directly affect the international grain supply in terms of security against famine, nor do they affect international market prices.

    The primary U. S. government mechanism for management of the farm economy is the Commodity Credit Corporation (CCC), which was created under legislation enacted in 1933. The CCC is a government corporation with independent spending authority managed under the Department of Agriculture. Under CCC supervision, a spectrum of programs was created designed to support farm income by controlling production and maintaining a grain reserve system as a mechanism for storing surplus foodstuffs, which could be released in periods of shortage.

35. Some analysts contend that farm support programs were themselves responsible for continually increased farm output, arguing that guaranteed minimum prices served to encourage production. See NACLA, "U. S. Grain Arsenal," p. 6.

36. Sanderson, "Food Fumble," p. 505.

37. Caldwell, _Food Crises and World Politics_, p. 23.

38. Hopkins and Puchala, "Perspectives on Food," p. 584.

39. Although $25-$30 billion is commonly accepted as the dollar amount of the food aid program, this figure should in no way be interpretd as the _cost_ of food aid. $25-$30 billion is what the food would have been worth had a commercial market existed for it, which it did not. Furthermore, had the food not been exported under PL 480, the cost of storing it would have been enormous. If it had been released on the international market, the value of these commodities would have dropped, thus making its total worth still smaller.

40. Lappe, _Food First_, p. 23.

41. Henry R. Nau, "The Diplomacy of World Food: Goals, Capabilities, Issues and Arenas," _International Organization_ 32, Summer 1978, p. 780.

42. _Ibid._

43. According to Wallensteen's analysis, the first and primary condition which must be present in order for a state to be able to utilize a commodity as a political instrument is scarcity. In addition, he believes that the price of a commodity, determined by supply and demand, may similarly reflect the potential effectiveness of the commodity as a source of political influence. If supply is short and demand high, a commodity is likely to be capable of demanding a high price in monetary terms, and thus may also command a high value politically. See Wallensteen, "Scarce Goods," p. 278.

44. Ibid.

45. Ibid.

46. Ibid., p. 283.

47. Some observers have raised the question as to whether the U. S.'s action independence with regard to food power may in any way be compromised by American reliance on agricultural imports - that is, can countries which are sources of U. S. food imports utilize these commodities as a means of retaliating, or deterring U. S. food power? According to statistics cited over almost a two decade period, U. S. agricultural imports as a share of total trade have declined significantly, from an average of more than 20 percent in the period 1958-1966, to a little more than 10 percent in 1974. See Ibid.

48. The evidence of food exports as a major earner of foreign exchange was dramatically illustrated in 1974. In that year, the United States experienced an overall trade deficit of more than $3 billion. Mineral and fuel imports accounted for a negative balance of more than $25 billion, against the major positive balances of manufactured products, which brought in approximately $7 billion in export earnings, and food exports, which in 1974 earned almost $14 billion abroad. See Ibid., p. 282.

49. Christensen, "Food and National Security," p. 302.

50. Wallensteen, "Scarce Goods," pp. 283-284.

51. U. S. Congress, Use of Food, p. 22.

52. Ibid., pp. 22-23.

53. Ibid., pp. 23-24.

54. Ibid., p. 24.

55. Ibid.

56. Ibid.

57. NACLA, "U. S. Grain Arsenal," p. 12.

58. Information on the structure of PL 480 is from Susan DeMarco and Susan Sechler, The Fields Have Turned Brown (Washington, D. C.: Agribusiness Accountability Project, 1975); NACLA, "U. S. Grain Arsenal"; and U. S. Congress, Use of Food.

59. U. S. Congress, Use of Food, p. 25.

60. Ibid.

61. Lappe, Food First, p. 329.

62. Ibid., pp. 330-331; NACLA, "U. S. Grain Arsenal," p. 13.

63. NACLA, "U. S. Grain Arsenal," p. 13.

64. Ibid.

65. Decisions on allocations under PL 480 are made by an assistant secretary level group called The Interagency Staff Committee, chaired by the Department of Agriculture and composed of representatives from the Departments of State, Commerce and Treasury, and National Security Council, the Council of Economic Advisors, and the Office of Management and Budget. Allocation decisions made by the Interagency Staff Committee are made in closed proceedings, and the Committee is not required to report its decisions to any committee of Congress.

66. NACLA, "U. S. Grain Arsenall, " pp. 13-14.

67. U. S. Congress, Use of Food, p. 25.

68. DeMarco and Sechler, The Fields Have Turned Brown, p. 41.

69. NACLA, "U. S. Grain Arsenal, p. 14.

70. Ibid., DeMarco and Sechler, The Fields Have Turned Brown, pp. 41-42.

71. NACLA, "U. S. Grain Arsenal," p. 15.

72. Ibid., p. 16.

73. Daniel E. Shaughnessy, "The Political Uses of Food Aid: Are Criteria Necessary?," in Peter G. Brown and Henry Shue (eds.), Food Policy (New York: Free Press, 1977), pp. 96-97.

74. U. S. Congress, Use of Food, p. 28.

75. Section 2 of Public Law 81-110 stated:

> that it is the policy of the United States to use export controls to the extent necessary (a) to protect the domestic economy from the excessive drain of scarce materials and to reduce the inflationary impact of abnormal foreign demand; (b) to further

the foreign policy of the United States and to aid in fulfilling its international responsibilities; and (c) to exercise the necessary vigilance over exports from the standpoint of their significance to the national security.

Under Section 3 the president was given authority to prohibit or limit exports, but was restricted from carrying out such measures with respect to agricultural commodities if they were not in a surplus condition, except if such actions were in the interest of U. S. foreign policy or the national security. Section 3(C) declared: the authority conferred by this section shall not be exercised with respect to any agricultural commodity, including fats and oils, during any period for which the supply of such commodity is determined by the Secretary of Agriculture to be in excess of the requirements of the domestic economy, except to the extent required to effectuate the policies set forth in clause (b) or clause (c) in Section 2 hereof.

See Schneider, Food, Foreign Policy, P. 9; U. S. Congress, "Use of Food," p. 67.

76. In 1962 the Export Control Act was amended to declare that U. S. policy would be to implement export controls in cooperation with all nations with which the United States had defense treaty commitments, and to formulate a unified policy to be observed by non-Communist nations in their dealings with Communist-dominated nations to further the national security and foreign policy objectives of the United States.

See U. S. Congress, Use of Food, p. 68.

Under the licensing system, several hundred commodities were eventually included on the general list for export to the Communist bloc. According to the amendment: The primary criteria for the denial or approval of a validated export license to Communist bloc nations was the degree to which the exported commodities: (1) contributed to the military or economic potential of the country; (2) would be directly applicable for military purposes; and (3) were available in other countries.

See Schneider, Food, Foreign Policy, p. 10.

77. Under Section 3 of Public Law 91-184, known as the Export Administration Act, the U. S. Congress delegated to the President broad authority to impose controls on American exports if it is necessary to protect the domestic economy, or if it is in the interest of the "foreign policy" or the "national security" of the United States. Section 3, article 2 of the act declares: It is the policy of the United Staes to use export controls (A) to the extent necessary to protect the domestic economy from excessive drain of scarce materials and to reduce the serious inflationary impact of abnormal foreign demand, (B) to the extent necessary to further significantly the foreign policy of the United States and to fulfill its international responsibilities,

and (C) to the extent necessary to exercise necessary vigilance over exports from the standpoint of their significance to the national security of the United States.

The Act further states that the United States shall seek the cooperation of its allies in implementing such controls. Article 3 states: It is the policy of the United States (A) to formulate, reformulate and, apply any necessary controls to the maximum extent possible in cooperation with all nations with which the United States has defense treaty commitments and (B) to formulate a unified trade control policy to be observed by all such nations.

Section 4(f) of the act restated the restriction on the utilization of export controls with respect to agricultural commodities: The authority of this section shall not be exercised with respect ot any agricultural commodity including fats and oils, during any period for which the supply of such a commodity is determined by the Secretary of Agriculture to be in excess of the requirements of the domestic economy, except to the extent required to effectuate the policies set forth in clause (B) or (C) of paragraph (2) of section 3 of this Act.

Thus, while certain restrictions are set forth with regard to export controls or agricultural commodities, such restrictions are not held to be applicable in cases perceived to be in the interest of United States foreign policy or national security. Other sections of the act stipulated penalties for violations, enforcement procedures, and reporting requirements. The Act has since been amended, but no significant changes have been instituted with regard to the President's authority to exercise such export controls as set forth in the 1969 act. See U. S. Congress, Use of Food, p. 69.

78. Schneider, Food, Foreign Policy, p. 10.

79. Weekly Compilation of Presidential Documents (Washington, D. C.: General Services Administration), pp. 32-33.

80. U. S. Congress, Use of Food, p. 71.

81. NACLA, "U. S. Grain Arsenal," p. 6.

82. Ibid., pp. 15-16.

83. Williams Report, pp. 283-284.

84. Ibid, p. 141.

CHAPTER 4

FOOD POLICY DECISIONS TOWARD THE SOVIET UNION: THE SOVIET GRAIN TRANSACTIONS, 1972; THE U. S. - SOVIET GRAIN AGREEMENT, 1975

This chapter looks at two cases of U. S. food policy decisions involving commercial sales of agricultural products to the Soviet Union. The first case examines the 1972 Soviet grain transactions, in which Soviet grain merchants purchased some 19 million tons of grain from American exporters, and which represented the emergence of the Soviets as a major grain customer. The Soviet purchases, which became known as the "great grain robbery," were made at bargain prices, heavily subsidized by American taxpayers. The sales contributed importantly to rapidly rising food prices in the United States, and to a worldwide food shortage that year. Although U. S. government officials later claimed they were simply outmaneuvered by the Soviets in the transactions, considerable evidence points to the fact that the sales were in large measure encouraged by the highest levels of the Nixon Administration, and were in fact part of a deliberate strategy to transform the entire international food system.

The second case examines the events and decisions leading up to the conclusion of the U. S. - Soviet grain agreement in 1975. Under the agreement, the United States promised to sell the Soviets up to eight million tons of wheat per year, in return for the Soviet's promise to purchase at least six million tons per year, for a period of five years.

The Soviet Grain Transactions, 1972

> We have chosen commercial sales of wheat to the Soviet Union over guarantees of an adequate diet for those impoverished Americans who subsist on surplus commodities. We have chosen, at least indirectly, to feed American livestock - in support of our taste for meat over grain - instead of meeting desperate human needs in West Africa, South Asia and elsewhere. We are forced to such results because we simply have no policy for choosing which needs to fill and which to ignore when we cannot fill them all.
>
> - Statement in August, 1973, by Senator George McGovern, former Food for Peace administrator under President John Kennedy.[1]

The period between the summer of 1972 and the fall of 1974 was perhaps the most dramatic two years in the history of United States foreign agricultural policy. In fiscal year 1972, U. S. farm exports reached a record \$8 billion, and by 1974 reached \$20 billion. Net farm income, which had remained between \$10 billion and \$14 billion for more than two decades, reached \$19 billion in 1972, and \$33 billion in 1973.[2] Food aid programs, which had accounted for as much as sixty percent of American agricultural exports during the previous two decades, were pushed aside, and support programs which had paid farmers not to plant acreage since the early 1950s were eliminated.

Within a span of a few months in mid-1972, the Soviets purchased from the United States more than 19 million tons of grain, worth approximately \$1.2 billion. This consisted of roughly 433 million bushels of wheat; 246 million bushels of corn; and 37 million bushels of soybeans. Soviet purchasers also made arrangements during this same period to buy 5 million tons of grain from Canada, 1 million tons each from France and Australia, and smaller amounts from Belgium, West Germany, and the Netherlands. These combined purchases accounted for the largest grain deal in history.[3]

The Soviets purchased this grain at $1.61 to $1.63 per bushel, or one cent to two cents below the prevailing market price. Shortly after the sales were completed, the price of wheat doubled, reaching a record price of more than $5 per bushel by August, 1973. Senior U. S. Department of Agriculture officials have always maintained that the entire episode of the Soviet purchases, including the jump in food prices, was a total surprise to them. Considerable evidence exists which indicates that not only were USDA officials fully aware of the events as they transpired, but that the Soviet sales were actually part of a deliberate policy strategy by members of the Nixon Administration.

Background to the Sales

The Soviets had made considerable progress in agriculture by devoting huge sums of money to increasing production under Khrushchev. Between 1953, the year of Stalin's death, and 1972, Soviet grain production doubled from 100 million to 200 million tons per year.[4] Soviet agriculture nevertheless remained highly inefficient, employing almost one-third of the labor force, and lagged considerably behind other industrialized countries. In addition to endemic structural problems, unfavorable weather has always been an important factor in Soviet agriculture. Two-thirds of Soviet grain production is in areas where rainfall is chronically undependable. Most of the production gains, moreover, had occurred in just four years prior to 1972, when these areas had experienced unusually moist, warmer weather.

Despite these gains, the Russian people still had less meat in their diets than people in most other developed countries, including the Communist states in neighboring Eastern Europe. In addition to a commitment to increase the amount of meat in the Soviet diet, the Russian leaders were also obligated to export approximately 8 million tons of grain to Eastern Europe, North Vietnam, North Korea, Cuba, and Egypt.

The inability to provide adequate diets to its people was clearly a major weakness in the Communist system. Clear evidence of this problem had been provided in late 1970, when violent riots occurred in Poland, when the government attempted to raise food prices. The Soviet leadership apparently made a decision to avoid such a situation by supplementing agricultural shortfalls with imports, some time following the Polish riots. According to Dan Morgan, "Whether the Polish riots were responsible for the Russian grain-buying spree is not known, but it seems likely that there was a connection."[5]

At the time when Soviet leaders made the commitment to improve the Soviet diet with more meat, it was unlikely that they believed this transition could be accomplished by increased domestic agricultural productivity. Soviet agriculture continued to suffer for a variety of reasons - poor transportation; inadequate incentives; and most particularly, the weather. For whatever gains might be made in agricultural efficiency and technology, a large percentage of Russia's arable land was still subject to generally unfavorable weather conditions.

Between 1968 and 1971, Soviet agricultural experts were successful in considerably expanding livestock herds through a 40 percent increase in the use of livestock feed. Particularly following the Polish workers' riots in 1970, the Soviet leadership placed unprecedented emphasis on "consumerism." By 1971, it became clear that these policies could only be continued by resorting to substantial grain purchases from the West.

Whether or not U. S. agricultural experts appreciated the significance of these developments in the Soviet Union, there is little doubt that Western grain merchants did. Continental, for example, maintained relations with Soviet grain traders throughout the 1960s. Continental president Michel Fribourg apparently believed all along that the Soviets would one day return as major customers for Western grain, and continued to support improved economic relations between the U. S. and Soviets.[6]

A major obstacle to expanded grain trade remained, however, in the 1963 Kennedy Administration guarantee to American unions that half of U. S. grain exports to the Soviet Union be carried in American vessels. Soon after the Nixon Administration assumed office, Continental began lobbying efforts to undo the 50-50 requirement. In fact, in December 1968, even before Nixon assumed office, Continental contracted to supply 400,000 tons of corn to the Soviets which, according to the contract, would come from the U. S. if the 50-50 stipulation was overturned before delivery.[7]

Soon after the new administration was in place, Continental approached new Assistant Secretary of Agriculture Clarence Palmby about revoking the 50-50 requirement, informing him that the Soviets were interested in purchasing American corn.[8] They indicated that while the initial purchase was to be for 300,000 tons, the Russians were additionally interested in buying U. S. corn, for cash, "on a continuing basis." Palmby passed on Continental's request to the new Secretary of Agriculture Clifford Hardin, who in turn wrote a memorandum to President Nixon on the matter. In the memorandum, Hardin stated his belief that the transaction "would serve the best interests of the United States," and urged revocation of the 50-50 requirement noting that it discriminated against only the Soviet Union, which he regarded as a major potential customer for American grain.[9]

Although Continental continued its lobbying efforts through the early months of the new administration, it failed to bring about a change in policy. In the meantime, however, Nixon Administration advisor on international economic policy Peter Peterson was planning what would become known as the New Economic Policy (NEP). By 1970, America's position in the international economy had reached a critical situation. The value of the U. S. dollar abroad had fallen dramatically; the U. S. balance of payments continued to deteriorate; and by 1971, the U. S. registered its first trade deficit of the twentieth century.

In May, 1970 Nixon appointed the Presidential Commission on International Trade and Investment Policy under the chairmanship of IBM's Albert L. Williams, in order to recommend major policy changes for a U. S. response to this situation.[10] A major component of the Williams Commission report was the laying of the groundwork for the purpose of expanding U. S. food exports by phasing out traditional U. S. farm programs. These programs had been designed to support farmer income and maintain low, stable food prices in the U. S., and were to now be replaced by a "free trade" market in agriculture, long favored by the grain merchants.

The first major administration move toward implementing the NEP and, in effect, laying the groundwork for the 1972 Soviet sales, was the announcement, on June 11, 1971 of the end to the licensing of grain exports to both China and the Soviet Union, and the removal of the 50-50 vessel requirement instituted by Kennedy.[12] The official change in policy would have little impact, however, without the cooperation of the labor unions, an area which became the subject of intense negotiations in the following months. Throughout this same period, contacts between Soviet trade officials and representatives of Continental and Cargill continued. In early November, union officials finally agreed to load vessels with Soviet-bound grain. Within weeks, Continental and Cargill announced initial sales of approximately 3 million tons of U. S. grain to the Soviet Union.[13]

The November, 1971 Soviet purchase of 3 million tons of U. S. grain should have served notice on U. S. agricultural officials of what was down the road in the way of Soviet plans. First of all, the Soviet purchase consisted not of wheat, which would have been purchased if their need was for increased supplies for direct consumption by people, but for corn, oats, barley, and sorghum, which were bought for animal feed (the Soviets had indicated that if such feed grain was not available, they would accept wheat, which could have also been used to feed livestock). The second important fact was that both 1970 and 1971 had been bumper years for Soviet grain crops. Thus Soviet officials had purchased 7.8 million tons of grain, even though these had been unusually good years.[14]

While economic considerations were obviously an extremely important factor in the U. S. decision to lift export controls to the Soviet Union, political considerations were not overlooked. One early Nixon-Kissinger foreign policy position had been the linking of technology sales to obtaining concessions in other areas. The administration had, for example, expressed a policy of refusing to expand trade with the Soviets until a broad range of issues had been settled, including settlement of $1.3 billion in World War II lend-lease payments still owed to the United States. One would have to expect, therefore, that if any major policy change was made toward the Soviets, that Kissinger, Nixon's chief foreign policy stategist, would have had a major voice in the decision. It was, in fact, national security adviser Kissinger who provided the first directive which indicated a change in U. S. policy toward selling food to the Soviet Union. On January 31, 1972, Kissinger issued a memorandum instructing the Departments of State, Commerce, and Agriculture on the importance of increased exports of agricultural commodities with the Soviets, including the granting of credit:

> One of the possible areas for increased trade with Russia relates to agricultural products and CCC (Commodity Credit Corporation) outlets. Agriculture should take the lead in a new public discussion. If negotiations with the Soviet Union should take place, the United States team should be headed by a representative of the Secretary of Agriculture.[15]

Kissinger issued a second memorandum on February 14, directing the Department of Agriculture, in cooperation with the other departments, to explore the possibilities of sales and extensions of credit to the Soviets. Although it was subsequently believed by most observers that the primary U. S. government interest in the transactions was economic in nature, the fact that the scenario began with instructions from the President's national security adviser draws immediate attention to the inter-relationship between foreign and economic policy in this issue. The directive read:

> The Department of Agriculture in cooperation with other interested agencies should take the lead in developing for the President's consideration a position and a negotiating scenario for handling the issue of grain sales to the USSR. This should include a recommendation on how the private transactions of the US grain sales would be related to Government actions including the US opening a CCC credit line and a Soviet commitment to draw on it. In cooperation with the Department of State, Agriculture should explore with the USSR the time and modalities of beginning such negotiations as soon as possible.[16]

There had been a number of indications by this period that the Soviet Union would be experiencing some degree of crop failure, which was not of course unusual. A report sent to USDA from its agricultural attache in Moscow on February 9, 1972 indicated that the Russian winter wheat crop had already "suffered significant damage" due to inadequate snow cover and extremely low temperatures. The attache concluded that the Soviets would probably have to purchase significant quantities of grain in the international market.[17]

Sould the Soviets have been forced to buy on the international market as USDA's experts believed as early as February, 1972, there was little mystery as to where this grain would have to come from. The United States, as reported in the February, 1972 USDA publication <u>Wheat Situation</u>, had more than 1.5 billion bushels of wheat stocks on hand. The rest of the world, however, was in short supply with the exception of Canada, which in February contracted to sell 3.5 million tons of wheat to the Soviets, with an option for the Soviets to purchase an additional 1.5 million tons to be shipped before the end of 1973.[18] Thus by February, 1972, the Canadian sale had already served to indicate an import need by the Soviets. In addition, the transaction exhausted the remainder of Canada's stocks, and by March both Canada and Australia had withdrawn from the world export market.

Early U. S. Preparations for the Sales

On February 25, 1972, a memo was drafted by Agriculture Department officials in response to Kissinger's February 14 request for a "negotiating scenario" on how to engineer grain sales to the Soviets. The memorandum was written by General Sales Manager of USDA's Export Marketing Service Clifford G. Pulvermacher, and approved by Assistant Secretary of Agriculture Clarence Palmby. As developed for Kissinger and presidential economic assistant Peter M. Flanigan, the scenario suggested approving a $500 million credit line to the Soviets in exchange for their commitment to purchase a certain amount of grain over a specified time period. These purchases were to be made through private trade channels.[19]

During March, 1972, Agriculture Secretary Butz accepted an invitation to visit the Soviet Union in April, to be accompanied by a U. S. negotiating team whose mission it would be to work out the details of a long-term agreement to buy American grain on credit. In early April, just prior to the Soviet trip, Butz sent a memo to President Nixon outlining USDA's objective to finance such an agreement through the Commodity Credit Corporation in return for a Soviet promise to regularize their purchases, as had been outlined in the earlier memo to Kissinger. The U. S. mission, accompanied by Secretary Butz, visited Moscow between April 8 and 18, led by Palmby and Pulvermacher. The major item discussed between the two sides involved the nature of the credit terms which the United States was willing to extend to the Soviets. Although the U. S. offer was reportedly quite generous, no agreement was reached.[20]

While in the Soviet Union, the group toured the Southern Ukraine region, a major growing area. The USDA team observed and reported clearly unfavorable crop conditions, which confirmed earlier reports by the U. S. agricultural attache stationed in Moscow. One Agriculture official along on the trip reached the conclusion, which was reported in The New York Times

> that the damage to the Soviet winter crops had been more severe than they had previously thought. Earlier estimates that 30 percent of the crop had been lost were now termed conservative.[21]

In addition, U. S. officials were aided by their knowledge that Soviet leaders had made a commitment to their people to improve their diets, which meant more meat. In order to accomplish this, the Soviets had begun to increase the size of their livestock herds. Furthermore, there were clear signs of an international food shortage in 1972, meaning that alternative soures of Soviet grain imports would be severely limited, or, more likely, nonexistent. According to one observer:

> All this the American officials were aware of - or at least they should have known. Besides, the Russians were talking of large sums, discussing a $750 million line of credit from the U. S. Government with as much as $500 million of it available in any one year.[22]

Additional reports by the attache on April 24 and April 25 warned Washington of the likelihood of a spring drought and unseasonably warm weather in major Russian growing regions. According to the attache: "This situation was not favorable to the normal germination and development of spring grains." The U. S. agricultural attache in Moscow continued to predict serious crop failures in the Soviet Union, sending cables on June 16, 1972, reporting spring grain to be "fair to poor"; and on June 26, estimating that "one-third of the winter grain acreage or 27 million acres was killed by winter weather conditions."[23]

All of these events should have, and despite their denials, undoubtedly did alert Nixon Administration Agriculture officials to the dimensions of the Soviet agricultural dilemma, and to the likelihood that the Soviets would be looking to the United States to fill their grain needs. In the May 8, 1972 issue of the farm journal Foreign Agriculture, Secretary Butz himself reported on the severe agricultural problems he had observed on his recent trip to the Soviet Union, and predicted that Soviet import needs might exceed previous estimates, although he failed to state by what amount:

> It might well be that we will be negotiating for annual sales in excess of $200 million worth of coarse grains and soybeans. This is based upon our best calculations - and frank discussions with General Secretary Brezhnev and Minister Matshevich - of the amount of grain Russia will need to boost her meat supply enough to keep the commitment made to the Russian people.[24]

On May 9, 1972 Assistant Secretary of Agriculture Palmby met with Soviet officials in Washington to renew discussion of possible Soviet grain purchases and reduced credit rates. This meeting was reported by Palmby in a memorandum sent to Secretary Butz, Secretary of Commerce Peterson, Peter Flanigan and Henry Kissinger.[25] And on May 18, 1972, Palmby sent Butz another memorandum, attaching a Memorandum of Understanding which outlined the basic elements of a proposed grain agreement with the Soviets, in which they would be granted and agree to purchase a minimum of $750 million worth of grain of Commodity Credit Corporation (CCC) credit. The Memorandum of Understanding was virtually identical to the actual agreement which would be announced some seven weeks later.[26]

According to testimony by Clarence Palmby, the decision to offer the Soviets a line of credit for purchasing U. S. grain was made by high level officials from USDA, the Council on International Economic Policy, and the National Security Council.

> The decision to offer a CCC line of credit to the Soviet Government against a commitment on their part to buy grain was made at the highest level in our Government.

In response to a question as to who was meant by "the highest level of our Government," Palmby replied:

> . . . The flow of paper was about like this: From Secretary Butz to the Counsellor for International Economic Policy, at that time held by Mr. Peter Peterson and then later by Peter Flanigan, and the National Security Council under Henry Kissinger, of course, was deeply involved.[27]

Under further questioning Palmby admitted that there were instructions from Deane Hinton of the White House Council on International Economic Policy "to close the deal as fast as possible." He further added what was his understanding of basic U. S. objectives in connection with the agreement:

> It was my impression at the time that the administration really wanted to capitalize on the ownership of agricultural commodities in this country to make progress on some other fronts in the whole foreign policy field.[28]

Further testimony by Palmby involved the question of whether the stocks of grain held by the Commodity Credit Corporation could have been sold directly to the Soviet Union, rather than selling the stocks to the private grain trading corporations, who then sold them to the Soviet Union. As a result of the grain traders' role, the CCC paid out approximately $300 million in export subsidies to the companies, which also, of course, additionally profited from the sales themselves.[29]

Mr. Palmby responded that the reason for involving the private corporations in the Russian sales was because of his (USDA's) belief that the U. S. should not engage in state trading. According to Palmby:

> This policy was adopted by Commodity Credit Corporation under several Presidents in several administrations, consistent with the language of the act and consistent with the guidance of the Congress.

Under further questioning, Palmby claimed that although he was in fact the official with responsibility for administering the CCC and the export subsidy program, he was not aware whether the CCC had legal authority to sell directly to the Soviet Union, if government officials had decided to do so, claiming that he nor any other officials had ever explored the possibility.[30]

The Soviet Grain Purchases

On June 27, 1972, USDA in Washington was notified that two groups of Soviet agricultural experts would be visiting the U. S., including the Soviet Foreign Trade minister - this information was withheld from public knowledge, but reported to the grain companies. Upon arriving in Washington, the Soviet mission initiated contact with Cargill and Continental. Two top Continental officials, Bernard Steinweg and Gregoire Ziv arrived in Washington from Paris to enter in negotiations with the Soviets, who reportedly stated their requirements at 4.5 million tons of wheat and 3 million tons of corn. Talks continued over the next week, joined by Fribourg, Continental's president who had urgently returned from a vacation in Spain; and by Palmby, who had just left his post with USDA. On July 5, the Soviets completed a deal with Continental to purchase 5.5 millions tons of wheat and 4.5 millions of corn.[31]

While these talks were occurring, other negotiations were taking place between the other half of the Soviet mission, and Assistant Secretary of Agriculture Carroll Brunthaver. In the interim Continental's Steigweg had sought, and received assurance from Brunthaver that the Department would continue to pay its wheat export subsidy. Thus USDA assured Continental of a guaranteed profit in its sale of approximately one-eight of the entire American wheat crop.[32]

The Soviet trade delegation had arrived in the United States on June 28 under a cloak of secrecy. It is unclear whether this secrecy was maintained at the request of the Soviets, or was devised by USDA. In any event, it was U. S. officials who chose to maintain the secretive nature of the Soviets' visit. While the private grain trading corporations traditionally operate in secrecy, the Soviets were here to arrange favorable credit terms with U. S. government officials, in addition to arranging actual purchases with private grain merchants. One group of Russians met with U. S. officials, while another group conducted meetings with grain exporters in New York, Minneapolis and Memphis. Within three weeks the Soviets had concluded arrangements for approximately two-thirds of their eventual purchase, and returned to Moscow.[33] USDA officials maintained that they were ignorant of the Soviet buying spree over these weeks, in which they contracted to buy approximately one-quarter of the U. S. wheat crop from the six major grain merchants, Continental Grain, Cargill, Dreyfus, Cook, Garnac and Bunge, with more than half of the total purchased from Continental. There is less disagreement over the fact that the public and wheat farmers were kept completely in the dark.

On July 8, 1972, President Nixon announced the conclusion of a grain agreement in which the Soviet Union had agreed to buy a minimum of $750 million of U. S. grain over a three year period. The credit would be supplied by the United States through the funds of the Commodity Credit Corporation and, according to the agreement, no more than $500 million in credit was to be outstanding at any time. While Nixon was announcing the agreement from San Clemente, Agriculture Secretary Butz and Commerce Secretary Peterson held a White House press conference, also attended by Assistant Secretary of Agriculture Carroll Brunthaver, who had by this time succeeded Palmby at USDA. Both officials were enthusiastic about the agreement, which Peterson referred to as "the largest agricultural transaction in history." He also stressed that the grain sales should be considered in view of the Nixon Administration's overall "strategic context," and were in the best interests of both countries.[34]

The government announcement made no mention of any Soviet buying activities, although Soviet trade representatives had already arranged to purchase approximately 8 million tons of grain for cash. Between July 10, just two days after the agreement was announced, and August 9, the Soviets repeatedly returned to the United States to purchase additional amounts of grain, which eventually would total some 12 million tons.[35]

Analysis

The grain trading corporations were in a position to profit enormously from information concerning Soviet buying activities. For one, Continental apparently had better inside knowledge or connections than the other grain merchants - Continental had already sold more than 8 million tons of grain before the other companies even learned of the Soviet presence in the U. S. Thus, Continental had the advantage of being able to buy grain futures before prices began to rise. If this information had been made public, American farmers would have had to be cut-in on the profits.

In addition, grain dealers with "inside information" from USDA could profit additionally by delaying the registration of their grain export sales with the Department after export subsidies were highest. (Those who knew what was occurring also knew that once the news of the massive Soviet purchases broke, prices would rise sharply, which would in turn lead to a significant increase in the export subsidy. Continental subsequently collected $.47 per bushel on more than 8 million tons of grain.) Those companies could thus make a profit on the Soviet grain deal two ways - first at the expense of the American farmers; and second at the expense of American taxpayers. It is estimated that these subsidies cost the U. S. Treasury about $316 million.[36]

While it is generally believed that all the grain merchants were implicated in these circumstances, two individuals and two firms seem to have been most intimately involved. The first case involves Clarence Palmby, Assistant Secretary of Agriculture under Butz until June 7, 1972, who was one of two high-level USDA officials to accompany the Secretary to the Soviet Union in April of that year. Palmby was approached by Continental with job offers at least three times - in January, February, and March 1972 - prior to his Soviet trip and his role in the negotiations with the Soviets. In fact, as early as March, Palmby had purchased a New York City condominium. Nevertheless, he denied contemplating accepting Continental's offers (their headquarters are in New York). On June 7, Palmby resigned from USDA and formally joined Continental one day later. These circumstances either point to a great deal of good luck, or, as it would seem to indicate, implicate USDA in a highly irregular relationship with at least one of the major grain merchants.

The other official who accompanied Butz to Moscow, general sales manager of the Department's export marketing service Clifford G. Pulvermacher, also seemed to be guilty of unethical conduct in the deal. Pulvermacher had announced, in March, his intention to retire from USDA effective June 30, 1972. Although the Bunge Corporation had been meeting with no success in attempting to conclude a deal with the Soviets for months, they finally reached agreement on August 2, one day after Pulvermacher officially joined Bunge.[37]

USDA conducted itself in suspicious ways beyond the direct involvement of Palmby and Pulvermacher. In general, USDA seemed to do all it could to assist the Soviets in maintaining the secretive nature of their activities in 1972. As late as July 31, after the Soviets had negotiated the purchase of more than 8 million tons of grain, USDA advised wheat farmers that the average price of wheat for that year would not exceed $1.31 per bushel. This enabled the grain dealers to purchase U. S. grain at highly favorable prices, fully knowing prices would rise considerably once knowledge of the Soviet purchases were made public.[38]

Thus a case can be made that USDA deliberately took actions to assist major grain trading corporations in making huge profits. This was not at the expense of the Soviets, who were able to make their original purchases at low prices, and make subsequent purchases at prices far below what they should have been because of USDA subsidies. U. S. officials should have had clear indications of Soviet intentions, particularly viz their second buying spree which began when they returned to the U. S. in late July. State Department officials, in fact, notified USDA's Foreign Agricultural Service that Exportkhleb (the official Soviet trade agency) officials had been issued visas to return to the U. S. in order to "negotiate with Continental Grain."[39]

As Soviet crop conditions deteriorated later that summer, Soviet officials were forced to return to the U. S. to conclude agreements for additional purchases. Despite the fact that their position was considerably more tenuous than on their initial visit, having obviously lost whatever advantage of "surprise" they might have had in June, Butz maintained export subsidies until the Soviets had purchased their remaining import needs. Only in late August, under pressure from the Office of Management and Budget, did Butz finally agree to suspend the export subsidy and allow the price of wheat to rise to its natural level.[40]

In addition, USDA officials took steps to courteously warn the grain merchants (including their former colleagues now in the companies' employ) of the coming suspension of subsidies, which served to allow the exporters to fill their purchases on the domestic market at low prices, and obtain their subsidies before prices rose and subsidies ended. Goldman reports that USDA officials notified by telephone ten of the largest grain exporters of the coming change in policy, which permitted the grain merchants to make last-minute purchases from U. S. farmers before the farmers and public were notified the following day. The

Department also allowed exporters five additional days to register for subsidies on sales made on or before August 24. Exporters took advantage of this opportunity to register for subsidy payments for 282 million bushels of wheat between August 24 and September 21, 167 million of which were sold to the Soviets. Thus the Soviets were able to purchase all of their import requirements at $1.61 - $1.63 per bushel. Three weeks later, the price of wheat had already reached $2.43 per bushel.[41]

Still another USDA official implicated in controversial action was Assistant Secretary of Agriculture Carroll Brunthaver, who left Cook Industries in order to replace Clarence Palmby.[42] Mr. Brunthaver has been blamed for withholding information on Soviet buying activities from U. S. farmers and the public, a charge which he has denied. Specifically, in testimony before the House Subcommittee on Livestock and Grain, he denied receiving knowledge of sales or subsidy information from Continental Grain.[43] Continental Grain officials, however, have sworn that they kept USDA officials closely informed of their dealings with the Soviets, and reported this information directly to Brunthaver. Senior Vice President Bernard Steinweg of Continental Grain has testified that he provided current sales information to Brunthaver, as well as indication that the Soviets planned to purchase at least an additional 7 million tons of grain in person on July 3, 1972. His testimoney has been corroborated by two other Continental Vice-Presidents who accompanied Steinweg on his visit. Officials from Cargill, Bunge, and Louis Dreyfus have also testified that they kept Brunghaver and USDA informed of their sales activities with the

Soviets.[44] Brunthaver and other USDA officials continued to deny that they knew the transactions were taking place.

A Wall Street Journal article on August 14, 1973, cites still other "coincidences" surrounding the participants in the Soviet grain transactions. These actions by officials with Continental, Dreyfus, and Garnac, each indicate efforts by the corporations to keep the Soviet sales a secret. Butz has conceded that USDA was aware of the dimensions of the Soviet crop failure through reports received on July 14 and August 18, both of which were withheld from public knowledge. It has further been documented that USDA received a report from its special representative in Moscow on July 5 which accurately predicted that the Soviet shortfall would amount to approximately 20 million tons.[46]

In his testimony on the circumstances surrounding the Soviet grain sales, Secretary Butz was quite explicit in his belief that the transactions needed to be put in a larger context than simply focusing on the motives for the Department of Agriculture's secretiveness; or conflicts of interest on the part of USDA officials; or profits made by the grain corporations; or the costs to the U. S. Treasury and American farmers and consumers. The Soviet grain deal, Butz stated, must be viewed for its value as an "historic development."

> To begin with, I think it is fair and even necessary to view the renewal of United States-U.S.S.R trade within the overall context of world political change. During the past 15 months, we have greatly strengthened the prospects for world peace. We have ended a tragic war. We have brought home the prisoners of war. We have eased tensions in the Middle East. We have achieved a detente with the Russians and rapproachement with the People's Republic of China. We have signed a significant agreement on arms limitation and a number of agreements for technical and economic cooperation with the U.S.S.R.
>
> I am quite willing to accept, for American agriculture, a major share of the credit for these history-making developments. Food has been a valuable tool in our strategy of peace - a lever that

> more than any other single factor has brought back into the world economy some 1.1 billion people - almost a third of the human race. Think of it![47]

The Failure to Expand Acreage

While a number of decisions, as well as non-decisions by Butz's USDA that permitted the Soviet Union to pull off what became known as "the great grain robbery" are likely to remain somewhat cloudy, one of the actions most difficult to justify was the failure to expand U. S. acreage for the 1973 crop year. Throughout 1972 Butz refused to lift acreage controls - despite full knowledge of the dimensions of the Soviet purchases - and did not act until he was ordered to do so in 1973.

The acreage management program for wheat announced to farmers on July 17, 1972 could have been justified if USDA was, despite considerable testimony to the contrary, genuinely ignorant of the Soviet wheat purchases which had already been contracted for a short time earlier. The announced policy of increasing acreage set-aside from 20 million to 25 million acres was, of course, designed to support farm prices by a traditionally acceptable method, particularly during an election year. That the program was not eliminated, or at least adjusted after the magnitude of the Soviet purchases was well-known goes beyond all logic, at least on the surface.

Despite denial of <u>prior</u> knowledge of the Soviet sales, USDA officials did report, albeit rather slowly, export developments between July and November of 1972.[48] The July 31 USDA publication <u>Wheat Situation</u> predicted a rise in exports of some 25 percent, "resulting in a moderate reduction in carryover" at the end of the crop year. In

September, Butz conceded before the House Agriculture Committee that wheat exports would be in the neighborhood of one billion bushels, "by far the largest total in our history." And the November Wheat Situation finally predicted that end of the year carryover stocks would be "the least since 1967."

Still, Butz failed to change the set-aside program. Some justifications have been cited, among them the obstacle of bureaucratic inertia, and the fact that farmers would have had to make adjustments in the midst of planning for planting. The most plausible explanation was probably Butz's continued priority of supporting higher farm prices, despite its contradiction with his commitment to a free-market economy, and despite the predictable effects such a policy would have on domestic food-price inflation.

Other government actors did attempt to get Butz to reconsider the set-aside program. The Nixon Administration was, after all, supposedly actively pursuing a major anti-inflation program. One such group was the Price Commission, established to administer Nixon's anti-inflation policies. The Price Commission pressed Butz to lift acreage restrictions prior to, and during the summer of 1972 but, according to one source

> It received in return the impression that (USDA) departmental policy was directed instead toward raising farm income and that, if higher food prices were the result, this was the commission's problem.

Although the Price Commission should have had some say in farm policy because of its direct implications for holding down inflation, it had little real power within the administration. In fact, the White House became particularly displeased by the Commission's actions which brought attention to rising food prices in an election year and it "in effect was instructed (by the White House) to take no further action on food."[49]

Secretary of the Treasury George Shultz reportedly attempted to convince Butz to reconsider the 1973 wheat acreage set-asides in light of the Soviet grain deal, as did the Cost of Living Council, but neither met with any success. According to Destler's analysis, Shultz, who had considerably more power in the administration than did the Price Commission, might have affected some change if he had chosen to push his case harder. Others who might have had more impact included Herbert Stein, Chairman of the Council of Economic Advisers, and Donald Rumsfeld, Director of the Cost of Living Council. Each were senior economic policy officials who should have been able to influence administration farm policy had they chosen to intervene. These officials were apparently not well-enough informed over the economic implications of the acreage set-aside program, or were simply not interested enough to attempt to intervene in the matter, which the powerful Secretary of Agriculture had clearly staked out for himself. Moreover, in the summer of 1972 the administration was preoccupied with the election campaign, and displaying dissension over the farm program, or even bringing it to the attention of the president, were probably additional factors. According to Destler:

> There is no reason to doubt that, for Richard Nixon that summer, farm policy was important mainly for its connection to farm votes. And that meant higher farm prices. Lifting wheat acreage restrictions would have dampened prices before the election, but the inflationary effects of not lifting them would not be felt until well after November 7.[50]

Some sector of the foreign policy community might have attempted to intervene on the basis of the effects that the set-aside program would have on decreased reserve stocks, which would mean substantially less food available for foreign policy purposes. But the most directly effected foreign policy sector was the development community, since developing countries were the most likely victims of both the decision to open up the Soviet market, and the reduction of food reserves and higher food prices. This group was never particularly influential in either foreign or foreign economic policy, however, and was unable to influence acreage decisions.[51]

In January 1973, just as the Nixon Administration was preparing to end its price control program, the December wholesale food price index was announced at a staggering 6 percent increase. The reason for the increase was, of course, not limited to farm policy decisions, but the justification for continued acreage restrictions in the face of grain scarcity became even more tenuous. Almost immediately, the Council of Economic Advisers ordered the easing of the set-aside program, which Butz finally agreed to, although some 7.4 million acres were still held out of production.[52]

The U. S. - Soviet Grain Agreement, 1975

In 1974 the Soviet grain crop was down to 184 million tons, as compared with the 1973 record crop of 212 million tons, and the Russians returned to the United States to buy food. In early October, Soviet grain buyers contracted to buy more than two million tons of corn, and nearly one million tons of wheat from their old friends at Continental and Cook. But unlike 1972, the companies were required to report the sales to the Department of Agriculture, which was under instructions to relay all such information to the White House.[53]

This time however, American grain supplies were not nearly as plentiful as they had been two years earlier - there had been a drought in the U. S. corn belt, and reserve stocks had been drastically depleted, due in large measure to the 1972 sales to the Soviets. Both corn and wheat prices were at near-record levels, and U. S. consumer food prices had still not been brought under control, and continued to be a major problem for the Ford Administration. U. S. food policy officials were determined that the events of 1972 not be repeated.

The day after being notified of the Cook and Continental contracts, on October 5, 1974, Treasury Secretary William Simon convened a meeting between company officials at the White House. With Secretary of Agriculture Butz in attendance, President Ford himself expressed his belief that the sales to the Soviets should be cancelled, citing "U. S. political pressure" and the possibility that the Congress might move to impose export controls. Two days later, on October 7, Butz announced the establishment of a "voluntary" system under which all grain sales over 50,000 tons to any foreign country would require prior approval by the Department of Agriculture.[54]

Over the course of the next few months, however, the U. S. supply situation eased and prices began to decline. On January 29, the Department of Agriculture announced the removal of the prior approval requirement for wheat and soybeans, and on March 6, for grain. Exports would still be monitored, and sales contracts had to be reported to the Department, but not until after they were signed. In the following months, predictions for the 1975 Soviet grain harvest steadily declined, and would, eventually, fall an additional 50 million tons to 133 million tons, or some 80 million tons below 1974. In mid-July, following Soviet purchases amounting to some 10 million metric tons over the course of one week, the Agriculture Department reinstituted a prior approval requirement for the grain companies, and requested the companies to withhold further sales until the August 11 corn estimate was reported. With the corn report indicating a decreased estimate, Secretary Butz announced continuation of the "temporary suspension" of sales to the Soviet Union pending results of the September crop report.[55]

One week later, amid mounting pressure by the AFL-CIO International Longshoreman's Association to demand higher shipping rates and increased use of American bottoms for the movement of grain bound for the Soviet Union, the union announced a boycott on loading the Soviet-bound grain. In making the announcement, AFL-CIO president George Meany also cited consumer fears of rising food prices, with the events of 1972 still freshly in mind. Following a series of meetings between labor leaders and high administration officials, including President Ford, Secretary of State Kissinger, Labor Secretary John Dunlop, O.M.B. Director James T. Lynn, and presidential economic advisor L. William Seidman, union officials agreed to postpone their boycott.[56] In return, the administration agreed to extend the moratorium on grain sales to the Soviets until mid-October. The administration also promised to negotiate an agreement with the Soviet Union to regularize Russian wheat purchases. To handle negotiations with the Soviets on this matter, President Ford established a new high-level food policy group, called the Food Committee of the Economic Policy Board and the National Security Council, chaired by the Secretaries of State and Treasury, and including representatives from Agriculture, Labor, Commerce, Chairmen of the CEA and CIEP, Director of O.M.B., and the presidential assistant for national security affairs. The Committee was also given "a continuing mandate to develop and maintain data on grain production and exports."[57]

When the U. S. delegation departed for Moscow on September 10, 1975, to begin negotiations with the Soviets on a long-term grain agreement, Under Secretary of State for Economic Affairs Charles Robinson headed the group. Thus the Department of Agriculture had by this time been placed in a largely subordinate role to the State Department on negotiating the pact, which was intended to regularize Russian grain purchases. This situation was, in effect, a complete reversal of 1972, when Agriculture was able to totally control U. S. agricultural trade with the Soviets. And when the administration asked Poland to temporarily suspend purchases of U. S. grain, to which it agreed, it was again a State Department official who made the request. Butz, in fact, had "bitterly opposed" the temporary embargo on grain sales to Poland.[58]

Reports surfaced at the time of the grain embargo stating the position that the 1975 temporary embargo on grain sales to the Soviet Union was politically motivated, and that Agriculture Secretary Butz did not control the decision. Instead, certain sources speculated, foreign policy officials had employed the embargo as a means of diplomatic leverage in order to keep the Soviets from disrupting the Sinai agreement between Israel and Egypt, which was engineered by Secretary of State Kissinger. In November, 1975, Dan Morgan of The Washington Post wrote:

> In the last two months, senior members of the Ford Administration have quietly stripped Agriculture Secretary Earl L. Butz of much of his power to formulate food policy and make grain export decisions on his own.
>
> As a result, some senior Agriculture Department officials are angry. They say that foreign agricultural policy is influenced

> more and more by 'instant experts' at the White House or 'striped pants' diplomats at the State Department.
>
> Top officials at other agencies confirm that the Agriculture Department has been required to share its decision-making power, and add that the power shift is necessary because of the large impact that food policy can have on prices at home and diplomacy abroad.[59]

Butz himself voiced the theory that the United States had used food as a "diplomatic tool" at the time of the grain agreement, and that this leverage had played a role in the Sinai Accord. In response to a question from Senator Clark before the Senate Subcommittee on Foreign Agricultural Policy in 1976, Butz replied:

> Senator Clark, frankly, I think we used it as a weapon when we stopped selling to the Soviets. I shouldn't say weapon. We used it as a diplomatic tool. We got an agreement out of it. I think we may have gotten more out of it. I couldn't prove it if my life depended on it. But I think everybody agrees that the Soviets could have stopped the Sinai agreement that Secretary Kissinger worked out between Egypt and Israel. They sat on the sidelines. I couldn't prove it. I am convinced they knew they had to come into our market for more grain. This is no time to rock the boat. You have to say we did, in fact, use food as a diplomatic tool. I think agripower will be more important than petropower.[60]

By 1975 the use of food as a diplomatic tool had already gained considerable credence in the Nixon and Ford Administrations: the withdrawal of food shipments to Chile beginning in 1970; the use of P.L. 480 dollars to support the war effort in Indochina between 1972 and 1974; and the granting of 100,000 tons of wheat to Egypt in 1974 in return for that country's cooperation in Henry Kissinger's peace plan. Kissinger in fact provides a common thread through each of the food-diplomacy activities. And while there is some evidence to suggest that Kissinger had developed a real concern for the world food situation because of its potential for international destabilization, Kissinger was by no means opposed to using food for specific political purposes.

The OPEC oil embargo had provided a difficult, but valuable lesson to U. S. foreign policy officials - the connection between OPEC's oil monopoly and America's food monopoly were obvious. In August 1974 the U. S. Central Intelligence Agency released a report titled "Potential Implications of Trends in World Population, Food Production, and Climate." The section titled "key Judgments" was blunt:

> The U. S. now provides nearly three-fourths of the world's net grain exports and its role is almost certain to grow over the next several decades. The world's increasing dependence on America's surpluses portends an increase in U. S. power and influence, especially vis-a-vis the food-deficit poor countries. Indeed, in times of shortage, the U. S. will face difficult choices about how to allocate its surplus between affluent purchasers and the hungry world.[61]

U. S. officials were also feeling threatened by the example which the OPEC embargo might set for other developing countries, which might jeopardize American political as well as economic interests, as put forth in documents such as the "New Development Strategy" and other aspects of the New International Economic Order. According to a NACLA analysis:

> Food has also become an important political weapon in U. S. efforts to counter growing challenges from the Third World. A principal concern of U. S. policymakers is the new drive by Third World commodity producers to assert control over their natural resources . . . . Increasingly, Third World countries are demanding not only control of their resources but also a basic restructuring of the international economic system that serves the interests of developed capitalist countries, particularly the United States, at the expense of the Third World.[62]

In early 1974 Kissinger and Treasury Secretary George Shultz had devised a strategy for the Washington Energy Conference, which attempted to utilize the food issue to "drive a wedge" between OPEC and the other developing countries. Kissinger apparently sought to convince the developing countries, caught in the dual price spirals of food and oil, that OPEC was to blame, and that it was the responsibility of the oil producers to help finance international food aid to the poor countries.[63] In a similar vein, President Ford himself attempted to raise the spectre of the U. S. food weapon, citing the possibility of Third World producer cartels in a speech before the U. N. General Assembly in the fall of 1974:

> The attempt by any nation to use one commodity for political purposes will inevitably tempt other countries to use their commodities for their own purposes. . . . It has not been our policy to use food as a political weapon despite the oil embargo and recent oil price and production decisions . . . . It would be tempting for the United States, beset by inflation and soaring energy prices, to turn a deaf ear to external appeals for food assistance.[64]

Consideration of the possibility of utilizing the Soviet Union's recent but considerable dependence on U. S. food for diplomatic purposes was a natural outgrowth of these developments. In early 1975 certain clues began to emerge that the Russians might be preparing to enter the grain market again. Russian representatives had been selling unusually large quantities of gold in Switzerland, and by June U. S. officials learned that the Russians had been secretly arranging to charter grain vessels on the Great Lakes. Finally, the Soviets suddenly expressed interest that they wished to participate in talks on a new international wheat agreement, which would give members priority access to an international grain pool. In addition, predictions on the size of the Soviet harvest continued to fall.

U. S. policy-makers began to discuss the possibility of gaining concessions from the Soviets in return for permission to purchase American grain. A group of officials gathered to examine the grain situation in the office of presidential adviser L. William Seidman, authorized by Kissinger and Under Secretary of State for Economic Affairs Charles Robinson. Others included in the group were Secretary of Labor John Dunlop; O.M.B. Director James Lynn; and Alan Greenspan and Paul McAvoy of the Council of Economic Advisers. Secretary of Agriculture Earl Butz was conspicuously absent from the group, not having been invited.[65]

The composition of the group was particularly interesting, besides the fact that Butz was not included, as it represented an alliance of foreign policy and domestic economic policy officials. With the exception of a compromise victory which Butz's Agriculture Department managed to retrieve on the U. S. position on the international grain reserve issue, Agriculture had been steadily losing control over U. S. food policy ever since the 1972 sales to the Soviet Union. And even on the international reserve issue, which had traditionally been Agriculture's exclusive domain, the decision was ultimately controlled by domestic economic policy officials.

As Secretary of Agriculture, Earl Butz ably represented the interests of corporate agribusiness, the objectives of which were best served by a free-market agricultural system, maximum exports, and the highest possible prices. This had brought Butz into direct conflict with domestic economic policy officials, who were still attempting to get a handle on U. S. inflation, in which skyrocketing food prices played a major part; and foreign policy officials, in particular Kissinger, who wished to utilize U. S. food resources for political purposes, which included making substantial amounts of food available on concessional terms. In late 1974 Butz had been overruled on two major food policy issues: in October, economic policy officials at Treasury and the Special Trade Representative's office had cancelled two large sales of U. S. wheat to the Soviet Union already contracted for with private corporations, while Butz was out of town. And in November, it was Secretary of State Kissinger who had delivered the opening address at the World Food Conference in Rome, with the Secretary of Agriculture in attendance.

Earl Butz was not, however, without his own base of support. The agricultural interests which Butz's Department of Agriculture represented accounted for some $200 billion worth of business in 1975. The Department's agricultural expertise, its network of attaches and other officials stationed around the world, and its close connections with the giant international agribusiness corporations constituted a major source of power in food issues. In the spring of 1975, USDA officials were the first U. S. government people to learn of the Soviet's actual intent to purchase American grain, having been notified by the grain companies. Assistant Secretary of Agriculture Richard Bell gave the companies the go-ahead to sell 10 million tons to the Soviets, without bothering to consult or even to inform any other U. S. officials.

Upon learning of the sales in July, both economic policy and foreign policy officials were chagrined. Members of the Economic Policy Board and the Council of Economic Advisers were caught by surprise, as Butz had successfully managed to control the decision process on a major food policy action without the potentially opposing actors even becoming involved. According to an analysis by Dan Morgan:

> The sales not only raised new concerns about inflation, but also deprived the Robinson group and the 'grain power' enthusiasts of any opportunity to act. The grain was contracted for and committed. The government's machinery for estimating Soviet grain requirements and for regulating the flow of grain out of the United States had malfunctioned.[66]

Still, the administration reacted slowly. It was not until eight days after the purchases had been made that exporters were requested to "voluntarily" suspend any new sales, and asked to report in advance any negotiations with the Russians. When, on August 11, the USDA crop report showed serious deterioration in the U. S. corn crop, the ten million tons had already been sold.

Where the administration had been reluctant to take any action after the July sales to the Soviets were revealed, the unions were not. George Meany had been a vocal critic of the soft line approach to the Soviets, and had little enthusiasm over grain sales which might result in fueling domestic inflation. Besides, as part of the 1972 U. S. - Soviet grain deal the Russians had promised to ship one-third of their American grain purchases in U. S. bottoms, and Meany believed they were cheating. The International Longshoremans Association, backed by the executive council of the AFL-CIO, threatened to boycott the loading of Soviet-bound grain unless the administration could guarantee that the sales would not affect domestic food prices.[67] On August 11 they began their boycott, and although they obeyed an order to return to work one week later, the union's presence on the issue considerably strengthened economic policy and foreign policy forces in the administration, which opposed Agriculture's policy of maximizing exports.[68]

The union's position thus coincided with the position held by both economic and foreign policy officials who advocated concluding a long-term agreement with the Soviets in order to regularize their erratic purchasing of American grain. Such an agreement would provide obvious benefits for controlling inflationary food prices, and was also in keeping with the foreign policy strategy for U. S. grain which had been developed by Kissinger and Robinson.

The logic behind the Kissinger-Robinson strategy to employ food power against the Soviet Union was simple. The Soviets were the world's largest oil producer, and they exported a considerable surplus, mostly to Eastern Europe, with a far smaller amount going to Western Europe. The Soviets were chronically short of grain, and had recently embarked on an ambitious five-year plan, a central feature of which was to increase the amount of meat in their citizens' diets. They were likely to be dependent on imported grain for the foreseeable future, at least. Furthermore, the Soviet government had consistent problems with maintaining adequate amounts of foreign exchange. The United States had grain that the Soviets needed, and the Soviets had oil that America wanted. Kissinger reportedly placed considerable value in obtaining an oil source outside of OPEC, and was apparently convinced that the Soviets needed U. S. grain so badly they would trade their oil for it, and would, moreover, even provide it at a discount.[69]

A number of additional factors contributed to make the Kissinger-Robinson grain for oil strategy even more attractive. Detente with the Soviets had been faltering, and conservative sectors were becoming increasingly critical of Ford Administration foreign policy (which was actually Nixon-Kissinger foreign policy) as being too soft on the Russians. The 1976 presidential elections were approaching and Ford moved to assume a harder foreign policy line, as the more conservative Ronald Reagan emerged as his principal challenger in the Republican party.[70]

Although a number of U. S. officials were apparently highly skeptical of the reality of the Kissinger-Robinson plan, they managed to enlist the support of Treasury Secretary William Simon - Simon later confessed that he thought the strategy was "ludicrous" - and persuaded President Ford to propose the plan to Leonid Brezhnev at the summit conference on European security at Helsinki in late July. Brezhnev reportedly expressed interest in the idea, but Ford apparently failed to mention the part about the oil discount.[71]

Robinson returned from Moscow on September 16, with the announcement that agreement on the long-term grain arrangement had been reached "in principle." No mention was made of any deal on Soviet oil, the strategy for which had remained secret from the public. Robinson then returned to Moscow on September 29 for a second round of negotiations. In the meantime, OPEC had raised the price of oil still higher, and the Kissinger-Robinson strategy reportedly had gained broader support among high White House officials. Agreement was smoothly reached on the grain arrangement, but the Soviets dismissed the oil accord plan completely. The longer the negotiations continued, and the embargo remained in place, however, the more persistent the administration became in pushing for it, despite the fact that each day Robinson cabled Washington informing the State Department that the Soviets had no intention of relenting.[72]

The plan continued to gather its own political momentum back in Washington, and in a press conference on October 10, Ford made the plan public, thereby deepening the commitment to come away with an oil deal even further. As the embargo on Soviet grain had no stated political purpose, having been established to protect domestic supplies for economic reasons, the grain companies remained free to provide the Soviets with grain from alternative sources, and continued to do so all through the embargo period. Cargill, Continental, Cook, Tradax, and

Louis Dreyfus were all in on the transactions, obtaining grain from Europe, Sweden, Romania, Canada, Australia, Argentina and Brazil.[73]

Finally, on October 20, 1975, following more than five weeks of negotiations, the U. S. - Soviet grain agreement was announced. Terms of the accord authorized the Soviets to purchase six to eight million tons of wheat and corn each year for the five years of the agreement. Larger amounts would be permitted if the United States agreed. No deal of any kind on oil was made.

President Ford had suffered severe political consequences over the prolonged embargo among important farm groups, just as the presidental election campaign was winding up, and a public loss of prestige over the failure to gain a concession from the Soviets on oil. In January, 1976, Ford declared that an American grain embargo against the Soviets in response to their activities in Angola would not even be considered, stating in a speech before the American Farm Bureau: "There is not the slightest doubt that if we tried to use grain for leverage, the Soviets could get along without American grain and ignore our views."[74] And on March 5, 1976, Ford announced that he was returning control of all agricultural policy, both domestic and foreign, to Agriculture Secretary Butz.

Chapter Notes

The Soviet Grain Transactions

1. Cited in Stephen Rosenfeld, "The Politics of Food," Foreign Policy (Spring 1974), p. 29.

2. I. M. Destler, Making Foreign Economic Policy (Washington, D. C.: Brookings Institution, 1980), p. 36.

3. U. S. Senate, Committee on Government Operations, Russian Grain Transactions, Hearings, July 20, 23, and 24, 1973 (Washington, D. C.: U. S. Government Printing Office, 1973), p. 218. (Hereafter referred to as "Grain Transaction Hearings.")

4. Dan Morgan, Merchants of Grain (New York: Viking Press, 1979), p. 10.

5. Ibid., p. 12.

6. Ibid., p. 140. Continental seemed quite adept at long-range planning, if USDA did not. During the 1960s, Continental had in training a young Soviet Exportkhleb agent in its Paris office, and it remained as its head of customer relations Russian emigre Gregoire Ziv.

7. Ibid., p. 141.

8. Palmby, who would play a central role in the 1972 Soviet wheat deal, was one of a cast of characters at USDA who had a long, intimate association with the grain industry. Between 1961 and his appointment to USDA, Palmby had served as executive vice-president of the U. S. Feed Grains Council, a major grain industry lobby which promoted the expansion of American grain exports. See NACLA, "U. S. Grain Arsenal," Latin American and Empire Report 9 (October 1975), p. 6.

9. Morgan, Merchants of Grain, p. 142.

10. Two major figures on the William's Commission with close ties to U. S. agribusiness were Edmund W. Littlefield, a member of the board of Del Monte Corporation; and William R. Pearce, Vice-President of Cargill, Inc., the world's largest grain merchant. See NACLA, "U. S. Grain Arsenal," p. 3.

11. Ibid., p. 4. See also Chapter 3, especially pp. 19-25.

12. According to the White House announcement, President Nixon had "decided to terminate the need to obtain Department of Commerce permission for the export of wheat, flour, and other grains, to China, Eastern Europe, and the Soviet Union." Weekly Compilation of Presidential Documents, June 14, 1971, p. 891.

13. Morgan, Merchants of Grain, p. 144; NACLA, "U. S. Grain Arsenal," p. 7. The evidence suggests that the grain corporations were amply represented at USDA during this period. In addition to Palmby, several of Palmby's subordinates at USDA had worked for the grain trade or other segments of agribusiness. In November, 1971, at about the time of the initial Soviet purchases from Continental and Cargill, Nixon appointed William Pearce, Cargill vice-president and prominent author of the Williams Commission Report, as deputy special trade representative with ambassadorial rank.

14. Morgan, Merchants of Grain, p. 144.

15. "Grain Transactions Hearings," p. 15.

16. Ibid.

17. Ibid. The report on damage to the Soviet winter wheat crop was confirmed the following day in the February 10 edition of Izvestia.

18. Ibid.

19. U. S. Senate, Committee on Government Operations, Russian Grain Transactions, Report No. 93-1003, (Washington, D. C.: U. S. Government Printing Office, 1974), p. 4. Hereafter referred to as "Grain Transactions Report." Despite U. S. eagerness to sell grain to the Soviets, and clear early indications that Soviet needs might be quite sizeable, Palmby denied in subsequent testimony that USDA officials were of the opinion that the Russian purchases would be as large as they were to be:

> I can say to you whether we were foolish or not foolish there was no one that thought they would buy that volume of grain, even the 750 million, (the amount the Soviets would be required to commit to purchase under the terms of the agreement) that quickly. But they bought considerably more than that. (See "Grain Transactions Hearings," p. 39).

20. "Grain Transactions Hearings," p. 17.

21. The New York Times, April 13, 1972, p. 11. Cited in Marshall I. Goldman, Detente and Dollars: Doing Business with the Soviets (New York: Basic Books, 1975).

22. William Robbins, The American Food Scandal: Why You Can't Eat Well on What You Earn (New York: William Morrow & Co., 1974), pp. 184-185.

23. "Grain Transactions Hearings," p. 17.

24. Ibid.

25. According to William Robbins, Soviet negotiators approached Palmby with a proposal for a government-to-government deal, in which the Soviets would purchase directly from government-held stocks. Although such a deal would have had several advantages over what actually occurred - the U. S. government would have known precisely how much the Soviets intended to purchase; it probably would have dampened price increases; and it would not have necessitated the U. S. government paying over $300 million in subsidies to private grain trading corporations - Palmby, Robbins reports, flatly rejected the Soviet proposal. Instead, he insisted, purchases would have to be channeled through the grain merchants. Less than one month later, he joined the corporation which accounted for nearly half the total Russian purchase. See Robbins, The American Food Scandal, pp. 185-186.

26. "Grain Transactions Hearings," p. 18.

27. Ibid., p. 31.

28. Ibid., p. 32.

29. Another impact of the grain sale was skyrocketing food prices in the U. S., which have been estimated to have cost the American consumer roughly $1 billion. Ibid., p. 49.

30. Ibid., pp. 47-48.

31. Robbins, The American Food Scandal, pp. 187-189.

32. Ibid., p. 190.

33. U. S. House of Representatives, Committee on Agriculture, Sale of Wheat to Russia, Hearings, September 14, 18 and 19, 1972 (Washington, D. C.: U. S. Government Printing Office, 1972), p. 218.

34. "Grain Transactions Hearings," p. 19.

35. Ibid., pp. 19-20.

36. Ibid., pp. 206-207; 210. For other accounts of the role of the grain corporations in the Soviet grain transactions, see Goldman, Detente and Dollars; Morgan, Merchants of Grain; and Robbins, The American Food Scandal.

37. "Sale of Wheat to Russia," pp. 284-285.

38. U. S. General Accounting Office, Russian Wheat Sales and Weaknesses in Agriculture's Management of Wheat Export Subsidy Program, Report to Congress (Washington, D. C.: GAO, 1973), p. 51. Hereafter referred to as "Russian Wheat Sales."

39. "Grain Transactions Hearings," pp. 20; 243.

40. Goldman, Detente and Dollars, p. 211.

41. "Russian Wheat Sales," pp. 16; 81.

42. In addition to Palmby, Pulvermacher and Brunthaver, other USDA officials in the revolving door with the grain merchants included Claude Merriman, who retired in early 1972 as assistant sales manager for commodity exports of USDA's Export Marketing Service (EMS), and became a consultant for Louis Dreyfus Corporation in June; George S. Shanklin, who replaced Merriman at EMS following seven years service as manager of the Washington, D. C. office of Bunge Corporation; and Charles Turnquist, who left Cargill in the summer of 1972 to become USDA Deputy Administrator for Commodity Operations. See A. V. Krebs, "Of the Grain Trade, By the Grain Trade, and For the Grain Trade," in Catherine Lerza and Michael Jacobson, Food for People Not for Profit (New York: Ballantine Books, 1975), pp. 356-357.

43. "Sale of Wheat to Russia," pp. 16; 75.

44. Bernard Steinweg, Senior Vice President Continental Grain, testified that he notified Assistant Secretary of Agriculture Carroll Brunthaver on July 3 that they had been contacted by the Soviets (on June 30) to purchase 4 million tons of wheat and 3 million tons of corn. Steinweg also testified that he contacted Brunthaver on July 6 and informed him of consummation of sales of four million tons of wheat. In testimony before the House Agriculture Committee, Secretary Butz nevertheless denied being notified of such activities at any time, a fact supported by Secretary Brunthaver. Steinweg confirmed his version of the events, which were corroborated by two colleagues at Continental, Samuel Sabin and James Good, in affidavits submitted for the record. Additional affidavits from officials of Cargill, Bunge, and Louis Dreyfus claimed that on various dates they similarly notified USDA officials of their planned transactions, and the quantities involved, with Soviet grain trade officials. Nevertheless, USDA failed to notify either American farmers or the U. S. public that such transactions were in process. These transactions were not, of course, incidental, making it highly unlikely that they were simply overlooked by Agriculture officials. They already constituted, in fact, the largest grain transaction in American history. Nevertheless, in repeated testimony, USDA officials denied knowledge of these facts. See "Grain Transactions Hearings," pp. 55-56; 70-76.

45. "Sale of Wheat to Russia," p. 11.

46. "Grain Transactions Hearings," p. 142.

47. Ibid., p. 92.

48. This account is based mainly on Destler, Foreign Economic Policy, pp. 44-48.

49. Ibid., p. 46.

50. Ibid., p. 47.

51. Ibid., p. 46.

52. Ibid., pp. 47-48.

The U. S. - Soviet Grain Agreement, 1975

53. Ibid., p. 103. In 1972, when there existed virtually no controls on U. S. exports, the Soviets purchased a total of about 20 million tons of grain.

54. Ibid., p. 104. Treasury Secretary Simon subsequently renegotiated the Soviet contracts, which, although they subsequently were allocated the same amount of grain, reversed the total amounts of corn and wheat.

55. Ibid., p. 105. The Soviets had hoped to harvest about 215 million metric tons. USDA estimated the Soviet harvest in preseason at 210 million metric tons.

56. Accounts differ as to whether Agriculture Secretary Butz was present at the September 9 meeting, at which the unions agreed to forestall their boycott. Destler says Butz was present; Dan Morgan contends he was not, and offers the following account of the meeting:

> On September 9, (Labor Secretary) Dunlop guided Meany and other labor leaders into the Oval Office at the White House for a meeting with President Ford. It was a political set piece. Presidential adviser Seidman read a prepared question: If the United States attempted to negotiate a permanent agreement that would establish ceilings and a regular procedure for future Soviet grain transactions, would the longshoremen load the ships? Silence. Dunlop nudged longshoremen President Teddy Gleason. 'Yeah,' said Gleason.

See Morgan, Merchants of Grain, p. 272. Morgan offers the most extensive account of the 1975 events.

57. U. S. Senate, Committee on Agriculture and Forestry, Subcommittee on Foreign Agricultural Policy, Who's Making Foreign Agricultural Policy?, Hearings, January 22 and 23, 1976 (Washington, D. C. : U. S. Government Printing Office, 1976), pp. 29-30.

58. Dan Morgan, "Butz Stripped of Powers on Farm and Food Policy," in Ross Talbott, The World Food Problem and U. S. Food Politics and Policies: 1972-1976 (Ames, Iowa: Iowa State University Press, 1977), p. 319. On September 10, the same day Robinson left for Moscow to begin negotiations on the grain agreement, the State Department uncovered that the grain companies were still selling grain to Poland. Since Poland was heavily dependent on the Soviet

Union for its grain supplies, the Polish sales were the equivalent of a leak in the embargo. The opposing factions once again squared off before President Ford in an Oval Office meeting. Secretary of Labor was again aligned with Kissinger, and despite sharp protests from Butz, the President agreed to extend the embargo to Poland.

59. Ibid.

60. U. S. Senate, Who's Making?, p. 19.

61. U. S. Central Intelligence Agency, Potential Implications of Trends in World Population, Food Production, and Climate (Washington, D. C.: C.I.A., 1974), p. 2.

62. NACLA, "U. S. Grain Arsenal," p. 4.

63. Leslie H. Gelb and Anthony Lake "Washington Dateline: Less Ford, More Politics," Foreign Policy 17 (Winter, 1974-75), pp. 179-180.

64. Cited in NACLA, "U. S. Grain Arsenal," p. 17.

65. Morgan, Merchants of Grain, p. 266.

66. Ibid., p. 268.

67. Butz later testified to the fact that he believed the union leader's motives were otherwise:

> In the meantime, we had the difficulty of loading boats and with the longshoremen refusing to load boats and with George Meany coming into the act and indicating they weren't going to load boats 'In the interest of protesting the cost of living in America.'
>
> This was a smoke screen. The real issue is how big a ripoff from the Treasury do they get. The smoke screen was: I am doing this to protect the cost of living in America. I may say what is sold.

U. S. Senate, Who's Making?, p. 8.

68. Ibid., pp. 268-269.

69. Ibid., p. 269.

70. One unintended consequence of the political difficulties generated by the embargo on grain sales to the Soviets during the negotiations appears to have been Ford's turning to Senator Robert Dole of Kansas as a runningmate. Dole was popular among Midwest farmers, but proved to be an overall political liability in the election.

71. Morgan, Merchants of Grain, p. 271.

72. *Ibid.*, P. 273.

73. *Ibid.*, p. 275. Oren Lee Staley, President of the National Farmers Organizations expressed similar displeasure with the highly visible role of Secretary of State Kissinger in the food policy process. He made it equally clear that farmers sharply opposed the 1975 grain agreement with the Soviet Union. Stanley stated:

> We not only don't know whom to trust - we don't even know who is making the decisions . . . .
>
> We may not agree with many of his policies and decisions, but we much prefer that the Secretary of Agriculture, whoever he may be, make such decisions as may be made to influence our grain export prices. We do not trust some bureaucrats in the State Department who only regards farmers as a source of materials he can use in the international game of influence and intrigue.

U. S. Senate, *Who' Making?*, pp. 105-106.

74. *Ibid.*, p. 279.

CHAPTER 5

U. S. FOREIGN ECONOMIC POLICY TOWARD THE ALLENDE REGIME IN CHILE, AND THE 1973 SOYBEAN EMBARGO

The two issues examined in this chapter represent nearly "opposite" cases of U. S. food policy decisions. The withdrawal of food aid from the government of Salvador Allende was part of an overall U. S. foreign policy strategy to destabilize the Chilean regime, and consideration of the possible domestic consequences of the policy was minimal. The decision to employ a soybean embargo in 1973 was, on the other hand, made for purely domestic reasons, and completely failed to consider the foreign policy implications of the action.

U. S. Foreign Economic Policy Toward the Allende Regime in Chile

The decision to withdraw food aid and other assistance from Chile was part of a comprehensive United States foreign economic policy strategy which was designed to destabilize the government of Salvador Allende Gossens. There is some disagreement as to whether the decision to withdraw economic assistance was made in retaliation for the Allende government's action to expropriate the considerable holdings of

American-based multinational corporations, or whether overall U. S. opposition to the existence of a socialist government in Latin America would have been sufficient cause for action, even without the expropriations. In fact United States policy had been directed at opposing the Allende regime well before the expropriations, but it is probably fair to say that the expropriation actions confirmed the fears of both the Nixon Administration and corporate officials, and provided direct justification for their activities.

The U. S. Central Intelligence Agency had actually opposed Allende's coming to power in Chile as early as 1958, when he first ran for president against Jorge Alessandri. Allende lost that election, and failed to win the presidency again in 1964, when he was defeated by Eduardo Frei. In 1970, however, Allende was finally elected President, winning with a plurality victory of 36 percent of the vote, and then being confirmed by the Chilean Congress.[1] The multinationals with large holdings in Chile had long feared for the security of their investments in the event of a leftist victory. As one analyst observed:

> The outcome of the Chilean elections sent shock waves through the private sector of the entire western world, but it raised particularly intense fears in the United States among private companies, such as the giant copper firms Anaconda and Kennecott, which had major capital assets invested in Chile. The multinational companies realized, however, that their large inventories and capital investments precluded the possibility of either a sudden withdrawal from the Chilean economy or an open confrontation with the new government.[2]

U. S. foreign-policy officials, particularly National Security Adviser Henry Kissinger, while undoubtedly aware of the economic implications of Allende's victory, also sought at this time to frame American concerns in traditional security terms. In a White House briefing on September 16, 1970, Kissinger spoke of the Chilean election, calling

Allende a "man backed by Communists, and probably a Communist himself . . . ." He further stated, in classic cold war rhetoric, his belief that

> A Communist government tends to be irreversible . . . .I have yet to meet somebody who firmly believes that if Allende wins there is likely to be another free election in Chile.

In fact, Allende was the leader of the Chilean Socialist Party, and had spent a large part of his political career opposing the Communist Party. This fact made the Nixon Administration's portrayal of the threat Allende posed even more surprising. Kissinger continued:

> Now it is fairly easy for one to predict that if Allende wins, there is a good chance that he will establish over a period of years some sort of Communist government. In that case you would have one not on an island off the coast which has not a traditional relationship and impact on Latin America, but in a major Latin American country you would have a Communist government, joining, for example, Argentina, which is already deeply divided, along a long frontier, joining Peru, which has already been heading in directions that have been difficult to deal with, and joining Bolivia, which has also gone in a more leftist, anti-U. S. direction, even without any of these developments.
>
> So I don't think we should delude ourselves that an Allende takeover in Chile would not present massive problems for us, and for democratic forces and for pro-U. S. forces in Latin America, and indeed to the whole Western Hemisphere.[3]

Although overall U. S. foreign policy toward Chile had involved opposing the election of Salvador Allende in both 1958 and 1964, when Allende was eventually elected in 1970, no coordinated offical policy existed. The State Department and the CIA were reportedly attempting to influence members of the Chilean Congress to vote against Allende in the confirmation process. The U. S. ambassador supported State's policy, but also remained in close contact with ITT officials, who favored a more aggressive policy aimed at immediately overthrowing Allende. The CIA also favored a coup, but took the course of establishing contacts with major banking and corporate interests in Chile, in the hope of fomenting economic difficulties, which they hoped would create the proper conditions for an eventual overthrow. The CIA also established contacts with the Chilean military which, they apparently concluded, was not inclined to step in to politics at that time.

ITT Director John McCone (who was, incidentally, a former head of the CIA) held a series of meetings with CIA Director Richard Helms between May and June, 1970, in anticipation of a possible Allende victory. Their strategy was reportedly to support either the Christian Democrat Radomiro Tomic or the Conservative Jorge Alessandri, who were Allende's major opponents. Sometime in June, Kissinger convened a White House meeting of the "committee of Forty," or "40 Committee," under his chairmanship, to consider what actions the United States would undertake if Allende was elected. The 40 Committee was the group responsible for approving covert Central Intelligence Agency activities. At that meeting, the group reportedly authorized the CIA to spend $400,000 for activities aimed at opposing Allende's election.[4]

The following month, the staff of the National Security Council began working on National Security Study Memorandum (NSSM) number 97, a top secret policy options paper on Chile. And on September 18, following Allende's plurality victory in the general election, but prior to the Chilean Congress' final decision between Allende and Alessandri, Kissinger reportedly recommended that the 40 Committee authorize the CIA to spend an additional $350,000 to bribe members of the Congress to vote for Alessandri. According to one investigative account, State Department officials "tacitly" admitted that the money was authorized, but denied that it was ever spent. Following Allende's confirmation as president by the Congress, the CIA was authorized to spend eight million dollars more between 1971 and 1973 for purposes of "destabilizing" the Allende government.[5]

Throughout Allende's reign, overall coordination of U. S. policy toward the Chilean regime was directed by the very highest levels of the Nixon Administration. Under White House direction, Kissinger and his N.S.C. staff acted to mediate the different approaches advocated by the various departments and agencies within the government, although the overall objective of opposing Allende was generally agreed upon. The degree of bureaucratic cooperation which characterized policy toward Allende appears to have been a somewhat rare occurrence in the annals of U. S. foreign policy. The Department of State was charged with handling public relations and propaganda activities, and Treasury with coordinating economic sanctions and the protection of American private investments, which it shared with State. The CIA, of course, was responsible for all intelligence-related activities. Even the Department of Defense got into the act, reportedly advancing the thesis that Allende also presented a strategic threat, which it had never before been considered, leading Kissinger to ironically characterize Chile as "a daggar pointed at the heart of Antarctica."[6]

Some difference in position apparently did exist with regard to approach between the Departments of State and Treasury, with Treasury advocating a somewhat harder line on economic sanctions. Treasury had by this time gained a high level of influence within the Nixon Administration, and had eclipsed State in certain areas of foreign policy, especially when economic interests were at stake. A number of factors contributed to the ascendance of the Treasury Department during this period. First was the influential positions occupied by both

Secretaries of the Treasury under Richard Nixon, George Shultz based on his personal authority and bureaucratic competence, and John Connally based on his close personal relationship with the President. The dominant influence of these two men was further accentuated by the relative weakness of Secretary of State William Rogers. State under Rogers was almost completely subordinated to Henry Kissinger and his staff at the National Security Council, due to both Roger's lack of power and Kissinger's overwhelming possession of it.

In addition, a fundamental change had occurred with the adoption of the Nixon Administration's New Economic Policy, in the greater responsibility assigned to the Treasury Department in the foreign policy decision-making process. A basic thrust of the NEP, as outlined in the William's Commission Report, was the recognition of the increased importance of the international economic dimension in foreign policy. Treasury's role was further increased during this period due to George Shultz' positions as both Secretary of the Treasury and Chairman of the cabinet-level Council on Economic Policy. As described by then White House press secretary Ronald Ziegler, Shultz became "The focal point and over-all coordinator of the entire economic policy decision-making process, both domestic and international."[7]

Treasury's role in foreign policy was particularly powerful with respect to the developing countries which had some degree of dependence on U. S. foreign aid, since Treasury was also vested with responsibility for formulating both bilateral and multilateral aid policy. Prior to the NEP, Treasury shared responsibility for policy regarding expropriations with State, which could be expected to have a "softer" line, since State's focus of concern was more balanced than Treasury's, which had more of a pro-business orientation. This had in fact been the case with respect to U. S. policy toward Peru's nationalization of the International Petroleum Corporation. Thus the realignment of foreign policy actors had a major impact on the resulting U. S. policy toward the Allende regime.

In January, 1971 President Nixon created the Council on International Economic Policy, and appointed Peter Peterson as director. Establishment of the CIEP represented the recognition by the Nixon Administration of the increasingly higher degree of interdependence between foreign policy and economic policy, both internationally and domestically. This realization that foreign policy was increasingly economic in content, and of the need for top-level focus in the government, meant a greater role for business' interest, in general, and the Department of the Treasury in particular in the foreign policy process. These structural changes were in keeping with Secretary of State Rogers' subsequent statement in a meeting with U. S. corporations with large interests in Chile that "the Nixon Administration was a 'business Administration' in favor of business and its mission was to protect business."[8] Members of the U. S. corporate community were, furthermore, adequately represented within the Treasury Department itself. Nearly all of the key officials had close ties with, and in most cases were former officers of, major U. S. banks and corporations.[9]

In July, 1971, the National Security Council issued NSSM number 131, which directed an interagency study on U. S. policy toward any countries which expropriated American-held interests. The groups which subsequently undertook the study included members of the Council on International Economic Policy, the Department of State, the Department of the Treasury, officials from the Overseas Private Investment Corporation (OPIC), the Export-Import Bank, National Security Adviser Henry Kissinger and Peter Peterson, assistant to the President for international economic policy.

While the United States was concerned with expropriations of U. S. interests other than those by Chile, the focus of concern at this time was clearly on Chile. It was reported that Kissinger and Peterson disagreed on what U. S. policy toward the Allende government's expropriations should be, with Kissinger taking a soft line, and Peterson supporting a hard line in order to make Chile an example for other countries which might be considering similar actions.[10]

The initial drafts for an official U. S. statement on investment and expropriation policy were prepared by both the State and Treasury Departments. The drafts reportedly reflected the difference in positions held by the two agencies, with Treasury expressing a harder line reflecting its pro-business orientation on foreign economic policy. Responsibility for preparing the final version was subsequently assigned to the Council on International Economic Policy. The new policy was announced by President Nixon on January 19, 1972:

> Thus, when a country expropriates a significant U. S. interest without making reasonable provision for such compensation to U. S. citizens, we will presume that the U. S. will not extend new bilateral economic benefits to the expropriating country unless and until it is determined that the country is taking reasonable steps to provide adequate compensation or that there are major factors affecting U. S. interests which require continuance of all or part of these benefits.
>
> In the face of the expropriatory circumstances just described, we will presume that the United States Government will withhold its support from loans under consideration in multilateral development banks . . . .
>
> The Departments of State, Treasury, and Commerce are increasing their interchange of views with the business community on

> problems relating to private U. S. investment abroad in order to improve government and business awareness of each other's concerns, actions and plans. The Department of State has set up a special office to follow expropriation cases in support of the Council on International Economic Policy.[11]

The official policy statement supposedly represented a melding of the views of State and Treasury, and the fact that six months had passed between the directive to review the policy, and its final release, seemed to indicate that considerable disagreement existed within the administration. Clearly, however, the hard line reflected in the Nixon announcement indicated a victory for Treasury's position over State. In fact, the statement was nearly identical to a detailed plan for U. S. strategy toward Chile sent to Henry Kissinger immediately following Allende's 1970 election, prepared by officials of the International Telephone and Telegraph Company:

> Inform President Allende that, if his policy requires expropriation of American property, the United States expects speedy compensation in U. S. dollars or convertible foreign currency as required by international law.
>
> Inform him that in the event speedy compensation is not forthcoming, there will be immediate repercussions in official and private circles. This could mean a stoppage of all loans by international banks and U. S. private banks . . . .
>
> Without informing President Allende, all U. S. aid funds already committed to Chile should be placed in the 'under review' status in order that entry of money into Chile is temporarily stopped with a view to a permanent cut-off if necessary. This includes 'funds in the pipeline' - letters of credit. . . .[12]

On the day of the announcement, Senator Edward Kennedy criticized the new expropriations policy, stating his belief that U. S. foreign policy should not be based on the interests of private investors:

> We should seek to divorce government policy from private interests. Our primary goals should be to encourage social and economic development of the low-income nations of the world as a fundamental prerequisite for political stability.[13]

Congressional opposition focused particularly on the provision which stated that the United States would, in the event of expropriations without resonable compensation, "withhold its support from loans under consideration in multilateral development banks."[14] This ran directly counter to the U. S. policy of channeling more aid through multilateral agencies, a direction which had been supported by the Nixon Administration. This opposition was rather short-lived, however, as the policy was subsequently formalized as legislation in the form of the Gonzalez Amendment, which the Nixon Administration actively supported. The amendment directed the president to instruct U. S. directors in the multilateral lending agencies to oppose the authorization of funds to any country which expropriated or nationalized American-owned properties. Thus, the dominant influence held by the United States in the multilateral lending institutions became an integral part of U. S. policy toward weakening the Allende Government in Chile.

The possibility of Allende coming to power had special significance for Nixon and Kissinger because of its context in laying the groundwork for detente with the Soviet Union and China. The anti-communist credentials which permitted Nixon and Kissinger to open up relations with the major established Communist countries made it imperative that they not allow "communism" to gain another foothold in Latin America. It mattered not at all that the Soviets had no role in Allende's election, nor that there was no evidence to suggest that Allende's victory would enhance their influence in Latin America. These were the men who had argued throughout their careers that all communist movements were "connected." Therefore, according to the logic which they themselves had espoused, Allende's election would jeopardize detente with both the Soviets and with China because it would be viewed by many as a Soviet-related communist incursion in the Western Hemisphere.

Still another factor which may have played a role in demanding hard-line opposition to the Allende government is what Richard Fagen calls "European dominoes." This refers to the possibility that the electoral victory of a Marxist would set an example for the powerful, yet perennially unsuccessful Socialist and Communist parties in Italy and France. The fear of "the electoral way to Marxist victory" was apparently felt in various segments of U. S. policy-making quarters since the beginning of the cold war. According to Fagen:

> The domino thinking undergirding this fear begins with a hypothesis about the demonstration effect of Chile: if the Chileans can do it, so the argument goes, then certainly Italian and French Socialists and Communists will have new life breathed into their historical vision - as will the Soviets. They will redouble their efforts and refine their tactics, learning from the successes and failures of Chile.[15]

There were also, of course, thoughts about a possible domino effect in Latin America itself, although not in the sense that the election of Allende would provide an example for other Socialist or Communist parties. In fact, no similar political configurations existed in Latin America, and very few countries held any sort of elections. The fear here was that Allende's economic nationalism might somehow inspire other Latin American governments to follow suit by nationalizing foreign interests and investment. This fear probably had more legitimacy than the others. The election would probably provide momentum to the rather loosely connected leftist movement throughout Latin America, which generally translated into anti-Americanism.

As mentioned, U. S. policy aimed at preventing a leftist victory and the election of Allende dated back to 1958. CIA Director William

Colby admitted that three million dollars in covert funds had gone to oppose Allende in 1964 (other estimates have placed the figure closer to $20 million). Between 1962 and 1965, the United States provided Chile with more than $600 million in direct economic assistance under the newly created Alliance for Progress, the highest per capital total in Latin America. Clearly, opposition to Allende and support of the Christian Democrats in Chile had become a deeply ingrained part of U. S. foreign policy well before Nixon and Kissinger took control of it is 1969. As summed up by Fagen:

> For a dozen years Chile had been considered a key country in Latin America, and Allende and the forces supporting him had been considered the chief threats to American interests. Directly and indirectly, perhaps a billion dollars in public funds had been committed by the United States during this period to the 'battle to preserve democracy in Chile,' largely defined as a battle to prevent the Left from coming to power. All of this, the investment of money, time, and prestige, was seen as in danger of going down the (Communist) drain.[16]

When Richard Nixon assumed the office of the President in 1969 he brought with him a well-established record of anti-Communism, particularly with respect to Latin America. Nixon was a hard-liner on Cuba, and had commissioned in 1969 the Rockefeller Report on the Americas, which was highly security-oriented, and fit comfortably with Nixon's position of seeking to protect U. S. corporate interests abroad, and as a friend of big business in general.

Both Nixon and National Security Adviser Kissinger seemed to have a particular interest in the events in Chile.[17] Although he was not at the time an official of the State Department, Kissinger personally chaired the special working group on Chile policy during the first three months of its existence in the Fall of 1970. The group was made up of members of State, Defense, Treasury, and the NSC, and would have normally been chaired by the Assistant Secretary of State for Inter-American Affairs, who at that time was Charles Meyer. This sort of direct personal involvement by the President himself and his chief foreign policy strategist effectively assured that policy would accurately reflect their personal objectives. The fact that the Treasury Department played the major bureaucratic role in U. S. policy toward Allende reflected not so much the power of the Department over the other agencies, such as State, as much as it was based on Nixon and Kissinger's desire that Treasury play that role. In other words, the structural changes which took place, such as the creation of the Council on International Economic Policy to coordinate policy, while not changing the important role played by Secretaries of the Treasury Schultz and Connally, in large part reflected the personal objectives of the President and his national security adviser.

Thus, while opposition to Allende was already well-established in the foreign policy establishment, the nationalization of the Anaconda and Kennecott copper mines by Allende apparently thrust Chile into the forefront of concern within the Treasury Department. The Administration moved quickly to implement its economic warfare "game plan" against Chile by cutting off Export-Import Bank credits; using its dominant influence in the World Bank and the Inter-American Development Bank to ensure that no further loans to Chile would be approved; pressuring private banks to cut off credit; and cutting back all forms of aid to Chile except military assistance.[18]

While American economic warfare directed against Chile was kept rather low-key in public, Nixon privately announced his intention "to make Chile's economy scream."[19] The termination of credit had a rather immediate impact on Chile's food supply. Food production was already down because of agrarian reform measures enacted during the previous administration of Eduardo Frei, and conservative landowners were themselves attempting to weaken the Allende government by limiting their agricultural production. Thus there are good indications that Chile would have experienced food problems even without outside interference. As a result of these factors, Chile's dependence on imported food doubled to $261 million in 1971, and grew to an estimated $283 million in 1972.

U. S.-directed economic policy had its impact by preventing the Allende government from replenishing its exhausted foreign exchange reserves from multilateral lending banks, and by cutting back bilateral economic assistance by some 85 percent. Title I food aid was terminated completely, although some Title II contributions were continued.

Unable to obtain needed credit to alleviate its food supply problems, and deprived of Title II food aid shipments, food grew increasingly scarce. Food prices began to rise sharply, which led to increased opposition to the regime by the Chilean middle class, the group generally believed to have been the cornerstone of Allende's political base. In what has been characterized as a "desperate" move, Allende even attempted to purchase U. S. wheat on a cash basis, which was also denied, "because of a political decision of the White House." Three days later, on September 11, 1973, the government of Salvador Allende was overthrown in a coup. Two weeks after the coup, the United States granted a credit sale of wheat to the military junta which had overthrown Allende in an amount which was eight times larger than the total allocation of food aid credits offered Chile during the entire three years Allende had been in power.[20] On September 26, the Department of Agriculture granted the junta a $24 million wheat credit, and two months later, a $28 million corn credit. In justifying the sudden turnaround in agricultural credits, Secretary of Agriculture Earl Butz conceded that

> they were made in the interest of national security. (The Allende government) was not friendly. It was essentially a diplomatic decision.[21]

It can be argued that the curtailment of U. S. food aid, and the denial of general economic credit by which the Allende government might have been able to purchase its food needs elsewhere did not in themselves cause the downfall of the government, nor even the severe food shortage problem. Clearly, domestic agricultural production problems, and the deliberate production cutback by the large landowners, contributed significantly. It is important, therefore, to understand U. S. food policy toward Chile as being but one part of an overall program of economic warfare, in which no one element can be singled out as being the cause of the weakening of the Allende government.

In response to a question regarding the importance of Chile obtaining sufficient capital for food imports, whether food imports were "central to a resolution of all other financial problems for Chile," Acting Assistant Secretary of State for Inter-American Affairs John H. Crimmins, stated in testimony in March, 1973:

> I think that this is in political terms the most sensitive of the imports for the Government; yes. Of course the availabilities around the world have tightened in the last year and a half because of the very massive purchases by the Soviet Union of many of the kinds of foodstuffs that Chile normally imports. I would agree that of all the imports the food imports are the most critical certainly from a political point of view in the Government.[22]

A great many variables apparently acted together in the Chilean case to bring about the downfall of the Allende government. No single policy action can be pinpointed, nor can it be proved that events would have been significantly altered without U. S. interference. The literature abounds with arguments which claim that the ineptness of the Allende government, and in particular its economic policies, brought about its own downfall. Such arguments cannot be proved any more easily.

The fact remains that certain U. S. actions were directed toward specific areas, such as attempting to affect the food supply, and that the sought after results were achieved, i.e., a food shortage did occur. It must also be realized, however, that a broad range of variables - the complementary activities of Chilean landowners; the effects of agrarian reform; cooperation from multilateral lending institutions; the character of the Chilean military; and the structure of the Chilean political, economic and social structure - all came together to produce the eventual result.

The most notable actor in the U. S. decision to cut off food assistance to the government of Salvador Allende was perhaps the one department which played no role at all, the U. S. Department of Agriculture. This can be explained by a number of factors. First, the financial investments of ITT, Anaconda and Kennecott far outweighed any possible commercial agricultural interests in Chile, which were relatively small. Second, the Allende regime was apparently perceived as a genuine threat to American ideological and corporate interests in the whole Western hemisphere, and as such was removed from the normal bureaucratic policy-making process. Moreover, food policy was only one component of overall U. S. strategy. And last, U. S. food policy at stake in Chile involved food assistance, and Agriculture was by this period turning its back on food assistance almost entirely, having handed over the area to Henry Kissinger for the possible political or strategic purposes it might serve.

The 1973 Soybean Embargo

The so-called "soybean phenomenon," can be traced back to the 1960s, when the crop emerged as a major success of green revolution technology.[23] Soybean production in the United States increased from approximately 300 million bushels in 1950; to 555 million bushels in 1960; to a record 1,271 million bushels by 1972.[24] The popularity of soybeans grew just as rapidly during this period, as general prosperity and an accompanying "upgrading" of diets in Western Europe and Japan created an increasingly larger demand for high protein animal feed.

American exports to meet that demand grew accordingly, from 287 million bushels in 1968-69 to 417 million bushels in 1971-72. Domestic carryover stocks decreased during this same period from 327 million bushels to 99 million bushels, as both exports and domestic consumption acted to drain supplies. Domestic production increases leveled off between 1968 and 1971, and although it increased slightly in 1972, reserve stocks declined even further to only 60 million bushels.[25]

In 1973 a number of circumstances converged to create a particularly tight international market situation for all oilseeds and protein meals, which put even further stress on the world's soybean supply. Aside from a relatively limited export capacity by Brazil, the world soybean market was an almost complete American monopoly. The United States produced more than 70 percent of all the soybeans grown in the world, and some countries were nearly totally dependent on American exports for their supplies. Japan imported some 97 percent of its soybean requirement, which it depended on as a major source of food and animal feed; 92 percent of these imports came from the United States.[26] Rapidly growing foreign demand and the limited capacity of the world to supply high protein feed combined to create a massive international shortage in 1973, equivalent to approximately 225 million bushels. American soybean exports were able to satisfy about half of this need, but the remaining shortage resulted in rapidly increasing prices, and scrambling by foreign buyers over any available supplies. Conditions were further exacerbated by unfavorable weather in the United States, which affected the 1972 harvest, and spring planting in 1973; and the Nixon Administration's devaluation of the dollar, which resulted in making U. S. exports more attaractive.[27]

U. S. food prices were still feeling the effects of the 1972 Soviet grain sales, and the rise in soybean prices, which had a direct impact on the cost of livestock production and was passed on to the American consumer in the form of higher meat prices, made matters that much worse. Between December, 1972, and August, 1973, the wholesale food price index jumped from 132.6 to 184.5.[28] The rise in consumer food prices was just as spectacular. Consumers were beginning to protest loudly, and American housewives were organizing to boycott meat.

In January, 1973 the Nixon Administration had lifted most price controls, and food price inflation had become a major policy issue. On

June 13, in an attempt to respond to the economic crisis, Nixon ordered a freeze on all prices with the exception of raw agricultural commodities. Two weeks later, on June 27, in order to protect domestic supplies, Secretary of Commerce Frederick B. Dent ordered a total embargo on the export of soybean, cottonseeds, and their derivative products. Although the embargo proved to be only temporary, the lasting impact of the action was considerable. An embargo of this sort had never been utilized in peacetime. American allies in Western Europe and Japan received little prior warning of the decision, which followed a pattern of independent action which had begun with Nixon's sudden economic rapproachement with the People's Republic of China in 1971. The credibility of U. S. foreign and international economic policy suffered serious damage as a result.[29]

The consequences of the 1972 soybean embargo were particularly distressing when it was subsequently learned that the export controls had not really been necessary. Six weeks after the embargo was imposed the Commerce Department announced that it had agreed to license fully 100 percent of soybeans contracted for export prior to June 13, although certain restrictions on derivative products remained in place. By September, all controls were lifted, and farmers began to harvest what proved to be a record soybean crop. U. S. exports for 1973 eventually topped the record level of the previous year, and all export contracts were fulfilled. Domestic supply problems cleared up, and prices returned to acceptable levels. These events led to serious questions concerning why the Nixon Administration had resorted to such extreme action in the face of what later proved to be spurious evidence.

In direct contrast to the course of action taken the previous year in the Russian grain transaction scenario, domestic economic policy interests outweighed foreign and foreign economic policy concerns in the decision to impose an embargo on soybean exports. The action was taken despite the fact that most Nixon Administration economic policy officials strongly opposed government controls on commercial economic transactions. A fundamental basis of the Administration's New Economic Policy was the principle of non-interference in trade and expansion of U. S. exports, particularly agricultural commodities.

When the Nixon Administration had imposed price controls in August, 1971, over the opposition of economic policy officials, it was widely believed that the step was taken in order to demonstrate firm economic policy action in the midst of the election campaign. Immediately following reelection, the administration moved to lift most wage and price controls. The results of the removal of the controls quickly proved disastrous, however, as domestic inflation, led by food prices, rose faster than at any time since the late 1940s.

Pressure for action to address the rapidly deteriorating domestic economic conditions quickly mounted from members of the Congress. Although Nixon ordered price ceilings imposed on beef, pork and lamb in late March, House and Senate members continued to press for an across-the-board price freeze, led by the powerful Chairman of the House Ways and Means Committee Wilbur Mills. When Cost of Living Council Executive Director John Dunlop moved to impose controls, or at least, a system of monitoring exports earlier in the month, he had encountered sharp administration resistance. Nevertheless, officials apparently recognized export controls as a likely outcome of a price freeze, considering the tight supply situation in agricultural commodities. As inflation continued to worsen, however, Dunlop finally succeeded, in May, in persuading administration officials that drastic action was necessary. Dunlop began to make preliminary plans to administer an overall freeze on retail food prices, and a group was formed to consider the imposition of export controls. Under the direction of Gary Seevers of the Council of Economic Advisers, the group included representatives from the Departments of Agriculture and Commerce, the Council on International Economic Policy, and the National Security Council. No member of the State Department was included in the group headed by Seevers.[30]

Although the group was, like the Administration itself, generally opposed to export controls, conditions convinced the group to recommend taking the first step toward implementing controls, which consisted of monitoring exports. George Shultz supported an export monitoring system as well. Thus, although still in the early stages of examining the issues, the Office of Export Administration, with assistance from the Department of Agriculture's Export Marketing Service, established an export reporting system for the purpose of monitoring exports of soybeans and other grains in tight supply.

Prior to the implementation of this system the government had no means of determining what volumes of commodities were being committed for export. The Department of Agriculture had certainly not served this function during the 1972 Soviet grain purchases, and appeared unwilling to do so in this case either. USDA under Earl Butz was firmly committed to a free-market, commercial agricultural economy with maximum exports and minimum government interference. The Office of Export Administration, which was charged with the task, was simply incapable of fulfilling it. According to an analysis by the U. S. General Accounting Office:

> The Office of Export Administration had never monitored massive quantities of oilseed and grain exports. It lacked adequate staffing, organizational resources, and an understanding of an experience in agricultural export activities; by itself, the Office could not perform the control function as directed by the President.[31]

Dunlop was by this time deeply concerned with food prices and inflation, and recommended retail food price controls and export controls on agricultural commodities as a solution. Just as economic policy officials became convinced of the need to take such serious measures in order to deal with what had come to be regarded as an economic crisis, Nixon political advisors such as John Connally, Melvin Laird and Bryce Harlow pushed for a similar course of action for political reasons. The political crisis know as Watergate apparently convinced Nixon to take the economic action which economic imperatives alone had failed to do.[32]

Nixon took this step when, on June 13, in a dramatic national television broadcast, he announced the imposition of a 60-day freeze on prices, except unprocessed agricultural products, to be followed by a "more effective system of controls." In his speech to the nation, Nixon cited export demand as a primary cause of the domestic economic crisis, and called for a new system for food export controls:

> One of the major reasons for the rise in food prices at home is that there is now an unprecedented demand abroad for the products of America's farms. Over the long run, increased food exports will be a vital factor in raising farm income, in improving our balance of payments, in supporting America's position of leadership in the world. In the short term, however, when we have shortages and sharply rising prices of food here at home, I have made this basic decision: In allocating the products of America's farms between markets abroad and those in the United States, we must put the American consumer first.[33]

Thus the decision to establish a new system of export controls on agricultural commodities was made in an effort to address domestic economic concerns. It also appeared that additional impetus for the action was provided by the Watergate-related political crisis. As Nixon's speech clearly pointed out, the administration was well aware that an embargo on food exports would have a negative impact on American efforts to increase food exports. And, as outlined in the NEP this in turn would hurt the administration's effort to utilize expanding agricultural exports to improve the U. S. balance of payments position. It would also undermine the use of trade as a source of supporting American foreign policy interests. As was vividly demonstrated in the Soviet grain deal, however, it was far less clear that increased farm exports would boost the income of American farmers; but amply clear that expanded exports would improve the income of corporate agribusiness.

The nature of the controls imposed, i.e. a total embargo on soybean exports, rather than some type of partial restrictions, as well as the suddenness with which it was imposed, left a number of unanswered questions. Although the President had committed himself to the use of export controls in the June 13 speech, he had promised to fulfill export commitments already made, and to consult with countries which would be affected by any actions taken. He did neither of these - the total embargo ordered cutting all contracts for soybean export signed prior to June 13 by 50 percent; and key State Department officials did not themselves learn of the embargo until minutes before the public announcement, and were not given the opportunity to notify foreign buyers.

Lack of experience, innacurate information, the overwhelming secretiveness of the grain trade and sheer incompetence could all have contributed to the decision to impose a total embargo, which, we viewed in hindsight, was ill-conceived. The structure of the decision-making process, however, which was tightly closed, and focused on one set of issues - the domestic economic and political crisis - to the exclusion of others, such a foreign policy and foreign economic policy considerations, no doubt had significant impact on the resulting policy decision. The 1974 G.A.O. analysis of the executive branch decision-making process on commodity shortage problems reached just such a conclusion:

> We found the existing executive branch short-supply decision-making process both fragmented and crisis-oriented. The Government's traditional reliance on the market system has restricted its ability to act promptly and effectively on an increasing number of short-supply situations.
>
> The use of 'crisis management' without effective communication, coordination, and planning has resulted in decisions that have been fragmented in terms of decision-making responsibility, application of alternative policy actions, sources and flows of policy priorities, options, and short-supply policy alternatives.[34]

One obvious area of question concerns the reasons why the administration failed to install mechanisms to monitor and control agricultural exports following the Soviet grain fiasco less than one year earlier. There apparently existed several sets of reasons why the Department of Agriculture failed to implement some sort of monitoring or reporting system, despite the enormous negative impact of the Russian grain purchases.

The Nixon Administration in general, and the agribusiness types who occupied the top positions in the Agriculture Department in particular, were said to be committed to a "free market" international economy. (This did not prvent them from utilizing agricultural exports to further U. S. political interests, so long as corporate agriculture continued to make its profit.) Maintenance of export controls was, in fact, antithetical to the interests of the agribusiness corporations which were more than adequately represented in USDA: it required reporting and maintaining just the sort of information that exporters prefer to keep secret. (As was well documented in the grain deal, USDA was more than willing to keep its information secret, however, not only from their business competitors, but from American farmers and the public alike.) Agriculture officials apparently felt that even the process of information-gathering was a dangerously close step toward export control.

It was not until food price inflation had become a major policy crisis and embarrassment that Nixon decided, in June, to utilize limited export controls to "put the American consumer first." Existing legislation in fact required any commodity for which export controls were levied be proved to be in short domestic supply and under serious demand for foreign export. Thus it was not until June 13 that the administration undertook a comprehensive survey of outstanding export contracts for soybeans.[35]

The manner in which the data were subsequently gathered proved to be highly unreliable. Assistant Secretary of Agriculture Bell had already encountered difficulty within the Department when he had attempted to install a voluntary registration system for agricultural export upon assuming his post three months earlier. Certain segments of USDA were skeptical that accurate figures would be maintained for the necessary reporting system, while otheres argued that the companies should be permitted to keep their information secret for competitive reasons. Furthermore, under the terms of the Export Administration Act, the Department of Commerce, not Agriculture, had responsibility for administering export controls, although with USDA's concurrence. Thus it was Commerce which sent out questionnaires to the companies, making the gathering of accurate information less reliable than had Agriculture acquired the information directly.

The information which came in showed a dramatic, alarming rise in the level of soybean and soybean meal exports, amounting to approximately twice the amount considered "available for export after fulfillment of domestic requirements."[36] USDA officials reportedly did not believe these readings were at all accurate, thinking they greatly exaggerated export demand. And although Secretary Butz would have opposed export controls, the credibility of Department of Agriculture officials to predict market conditions had eroded so severely that they were in no position to oppose the controls. According to Destler's analysis:

> The credibility of the 'expert' agency thus evaporated just at the time that expertise was most important as a check on the alarming, but misleading, data the government had just gathered. So Secretary Butz reluctantly endorsed the embargo order, and an ignorant government took unnecessary, damaging export control action.[37]

That foreign policy considerations were subordinated to domestic economic and political concerns in the decision to impose the 1973 soybean embargo could, at least to some degree, be attributed to the decision-making process itself. This conclusion is not intended to suggest that had the decision-making process been otherwise structured to take into account foreign policy concerns, no embargo would have been imposed. Domestic pressures were very real, and it is of course not possible to prove a scenario which never occurred. Still the contrast between the decision to arrange the Soviet grain purchases only a year earlier, in which foreign policy actors and concerns were clearly significant; and a decision made with minimal involvement of foreign-policy actors, is more than a little interesting.

The formal stucture for economic policy decision-making in the second Nixon Administration was the Council on Economic Policy (CEP) under George Shultz. The role of the CEP was to coordinate overall economic policy, domestic and foreign, and it was this group which was responsible for making the decision to impose export controls in 1973. Shultz had designated an interagency task force, in which the Department of State participated, to explore control systems, legal issues, and the need to consult with importing countries.

During this same period, however, a far smaller, restricted, ad hoc group under Seevers was also considering the question of imposing export controls. And it was this group, which consisted essentially of the Cost of Living Council and the Council of Economic Advisers, which controlled the decision. The focus, or mission of this group, as a result of its composition, was naturally on domestic economic issues. Their major concern was to insure an adequate supply of soybeans for domestic needs, both for livestock consumption and to keep the soybean crushing industry operating at maximum capacity.

Acting on the basis of these concerns, and supported by the recently gathered statistics on supply and export demand compiled by the Department of Commerce (which subsequently proved to be grossly inaccurate), the group made its decision. That the decision to impose the embargo failed to consider the foreign policy implications of such an action was predictable.

The Congress wanted food prices controlled in order to stem inflation, a concern shared by the Cost of Living Council. White House political advisers believed the President needed to take some form of decisive action to demonstrate the administration's ability to be decisive despite the problems generated by Watergate. Those groups which would have represented international interests were never admitted into the decision process. The State Department was not included in either the CEP or the Seevers group. And the NSC representative on the Seevers group was designated only to act as a "conduit" of information to Kissinger, and was, furthermore, new to his job.[38] The State Department was later called in to deal with the highly negative reaction generated by the embargo in Western Europe and Japan, because no consultation or even advance notice was given to major U. S. allies. Another factor may have been the upcoming Nixon-Brezhnev summit, which may have acted to focus U. S. foreign policy concerns unduly on the Soviet Union.[39] Consideration should also be given to the overall low level of importance which the Department of State was accorded under the Nixon Administration. Under Nixon, it was Henry Kissinger and his staff at the National Security Council which played the major role in high-level foreign policy decisions. The decision process which led to the 1973 soybean embargo was conducted at the highest level of government, with no opportunity given to senior staff level bureaucrats at the State Department, who might have injected a broader range of considerations into the discussions, and would have had a better understanding of the implications of the decisions for foreign policy.

There may exist some disagreement as to whether domestic economic concerns or domestic political pressures were the primary determinant in the decision to impose export controls. Certainly it was some combination of the two, to the exclusion of any foreign policy considerations. According to the analysis by Edward Graziano in the 1976 Report of the Commission on the Organization of the Government for the Conduct of Foreign Policy:

> The possibility of starvation among foreign countries dependent on America as a reliable supplier was never really considered in the decision process. The fact that export controls could have caused a change in other countries' foreign policies toward the United States was never considered. No one considered international notification of, and consultation on, the imposition of export controls. Quite possibly, the creation of diplomatic formalities, to provide a minimum of buyer notification, would have helped to diffuse foreign reaction.[40]

In addition to the fact that certain steps could have been taken to soften the impact of the soybean embargo on importing states, consideration of the foreign policy and foreign economic policy consequences of the embargo would have certainly provided some counterweight to the domestic considerations upon which the decision was made. Even if such considerations had failed to alter the eventual decision, it might have acted to at least delay a hastily-made decision, which quickly revealed itself to have been unnecessary in the first place.

It must also be noted that in addition to the negative impact the soybean embargo had on U. S. foreign policy and foreign economic policy interests, it also failed to achieve its domestic objectives, which were to hold down food prices and avoid a shortage at home. It did not act to reduce food prices, even for soybeans, most economists agreeing that prices were largely unaffected by the action. And with respect to alleviating a domestic supply shortage, it was soon revealed that no such shortage had ever existed in the first place, and that the data which had led officials to that conclusion was simply inaccurate.

Chapter Notes

U. S. Food Policy Toward the Allende Regime

1. Richard R. Fagen, "The United States and Chile: Roots and Branches," Foreign Affairs 53 (January, 1975), p. 297; Mitchel B. Wallerstein, Food For War - Food For Peace (Cambridge, Massachusetts: MIT Press, 1980), p. 256.

2. Wallerstein, Food For War, pp. 156-157.

3. U. S. Senate, Committee on Foreign Relations, Multinational Corporations and United States Foreign Policy, Hearings, March and April, 1973 (Washington, D. C.: U. S. Government Printing Office, 1973), p. 543. At the time of this briefing, Allende had not yet been formally elected. Since Allende had won a plurality but not a majority, Chilean election law passed the decision to the Congress, which then decided between the two highest candidates. Although that decision was not scheduled until October 24, the normal practice was for the Congress to elect the candidate who had received the highest number of votes.

4. Fagen, "The United States and Chile," pp. 297-298; James Petras and Morris Morley, The United States and Chile: Imperialism and the Overthrow of the Allende Government (New York: Monthly Review Press, 1975), pp. 27-28.

5. Fagen, "The United States and Chile," p. 298. It has been pointed out that considering inflation and the dollar black market which existed in Chile, the amount was probably worth between 40 and 50 million dollars.

6. Ibid., pp. 304-305.

7. Elizabeth Farnsworth, Richard Feinberg, and Eric Leenson, "Facing the Blockage," NACLA's Latin America and Empire Report 7 (January, 1973), p. 11.

8. Petras, The United States and Chile, p. 89.

9. For a listing of the most prominent officials in the formulation of Nixon economic policy and their associations with U. S. banks and corporations, see Farnsworth, "Facing the Blockade," p. 14. This issue also contains a thorough review of the total measures of the so-called "invisible blockade."

10. Ibid., p. 12.

11. Press release, "Policy Statement, Economic Assistance and Investment Security in Developing Nations," January 19, 1972, reprinted in Ibid., p. 13.

12. U. S. Senate, Multinational Corporations, pp. 720-721.

13. Congressional Record, January 4, 1972, p. S. 317.

14. Farnsworth, "Facing the Blockade," p. 13.

15. Fagen, "The United States and Chile," p. 302.

16. Ibid., p. 304.

17. See U. S. Senate, "Multinational Corporations," p. 701 for an account of this matter.

18. Richard J. Barnet and Ronald E. Miller, Global Reach (New York: Simon and Schuster), p. 83.

19. Wallerstein, Food For War, p. 158.

20. Stephen Rosenfeld, "The Politics of Food," Foreign Policy (Spring, 1974), p. 22.

21. Farnsworth, "Facing the Blockade," p. 13.

22. U. S. House of Representatives, Subcommittee on Inter-American Affairs, United States and Chile During the Allende Year, 1970-1973, Briefing, March 6, 1973 (Washington, D. C.: U. S. Government Printing Office, 1973), p. 88.

The 1973 Soybean Embargo

23. The description of events in this section is based primarily on accounts by I. M. Destler, Making Foreign Economic Policy (Washington, D. C.: The Brookings Institution, 1980) Chapter 5; and Edward Graziano, "Commodity Export Controls: The Soybean Case, 1973," in Edward K. Hamilton, "Cases on a Decade of U. S. Foreign Economic Policy: 1965-74," in Report of the Commission on the Organization of the Government for the Conduct of Foreign Policy ("Murphy Commission") (Washington, D. C.: U. S. Government Printing Office, 1976), Appendix H, pp. 18-32.

24. Destler, Foreign Economic Policy, p. 51.

25. Graziano, "Commodity Controls," p. 18.

26. Destler, Foreign Economic Policy, p. 51.

27. Graziano, "Commodity Controls," p. 19.

28. Destler, Foreign Economic Policy, pp. 50-51.

29. Ibid., pp. 52-53.

30. Graziano, "Commodity Controls," p. 19. The Secretary of Commerce has traditionally been the primary decision-maker in U. S. export control policy. As the economic dimensions of tight-supply situations have become more complex however, and have larger implications for broader aspects of both economic and foreign policy, more actors have become involved. For some time the United States followed an "incrementalist export policy" on most commodities, based on the long-range objective of improving the balance of payments position. In the early 1970s, however, certain short-range issues began to receive attention, particularly cases where export policies might be contributing to food price inflation, as appeared to be the case with the tight-supply situation of soybeans. See U. S. General Accounting Office, _U. S. Actions Needed to Cope With Commodity Shortages_ (Washington, D. C.: U. S. Government Accounting Office, 1974), p. 24.

31. GAO, _U. S. Actions_, p. 31.

32. Destler, _Foreign Economic Policy_, p. 58.

33. "Address to the Nation Announcing Price Control Measures," cited in _Ibid._, p. 59.

34. GAO, _U. S. Actions_, p. 17.

35. Destler, _Foreign Economic Policy_, pp. 54-55.

36. _Ibid._, p. 55.

37. _Ibid._, p. 56. While subsequent evidence confirmed otherwise, USDA officials nevertheless had blamed their own misreadings of demand conditions on the Soviet grain deal fiasco less than a year earlier. And throughout the early part of 1973, Agriculture officials had repeatedly grossly underestimated food price inflation to the extent that their estimates became laughable. In January, for example, officials estimated food price inflation for 1973 at 6 - 6.5 percent. One week later, figures for January alone showed an increase of some 2 - 3 percent.

38. Graziano, "Commodity Controls," p. 20.

39. _Ibid._, p. 21.

40. _Ibid._, p. 24.

CHAPTER 6

U. S. FOOD POLICY TOWARD THE THIRD WORLD, 1974-75

The two cases examined in this chapter illustrate the emergence of the food issue as a major factor in United States foreign policy toward the Third World during the mid-1970s. By 1974 the U. S. food assistance program had changed dramatically. PL 480 had accomplished its purpose of disposing of surplus food and developing new markets, and was being phased out by USDA. What was left of the program was seized by national security officials, however, and for a few years served as a major instrument in support of U. S. foreign policy objectives in the Third World.

Despite Henry Kissinger's overt use of U. S. food for such purposes, as the new Secretary of State he played an instrumental role in the convening of the 1974 World Food Conference in Rome, and genuinely attempted to commit the United States to increased levels of food assistance to needy countries, a move which was opposed by Agriculture Department officials. Economic officials also emerged at this time to contest the foreign policy uses of food because of the impact such activities had on the domestic economy.

Although elimination of an American-held grain reserve was a fundamental part of the transition to a commercial agriculture system following 1972, by 1974 Secretary of State Kissinger was attempting to pledge the United States to an international effort to rebuild food reserves. Again, USDA opposed Kissinger's position.

The Use of Food Resources in U. S. Foreign Policy Toward the Third World, 1974 - 75

Although the Food for Peace Program had existed since 1954, it was not until 1973 that the program became the target of significant competing interests within the U. S. government. Up until that time, decisions made under Public Law 480 were more or less "automatic." The interests of the major departments with stakes in the program, namely the Departments of State and Agriculture, had fundamentally complementary sets of objectives. The Secretary of Agriculture would determine the amounts of surplus food to be distributed under the various titles of the program, and the Department of State and AID would have the major voices in which countries received allocations, based on a balancing of political, diplomatic, and humanitarian objectives.

In 1973, food policy emerged as a major political issue as world hunger, food reserves, domestic food price inflation, the U. S. balance of payments and the use of food aid in support of continuing U. S. operations in Southeast Asia all acted to inject a new level of controversy into the area. Consequently, a whole new group of government departments and agencies suddenly developed their own stakes in food policy decisions, the most prominent of which were the Treasury Department, the Office of Management and Budget (OMB), the Council of Economic Advisers (CEA), and the Council on International Economic Policy (CIEP).[1]

To complicate matters further, the sets of objectives which had traditionally been represented by State and Agriculture changed as well. With Congress moving to cut off the administration's funding for its operations in Indochina, PL 480 became an increasingly important source of revenue for the war, and AID's humanitarian and development objectives were cast aside. Agriculture, which had traditionally represented the interests of farmers and, to a lesser extent, the American consumer, had changed dramatically in complexion and now represented the interests of U. S.-based corporate agribusiness.[2] According to Leslie Gelb and Anthony Lake:

> Thus, instead of a USDA-State/AID hand-holding operation, there was now a schizophrenic USDA and a desperate State Department, which had to share power with four other parts of the executive branch. There were hard choices to be made: the more food aid, the more inflation; the less food aid, the more starvation. And food aid was becoming a public - if little understood - issue, as headlines about the Sahel, India, and Bangladesh became more dramatic. The politics of food aid had begun in earnest.[3]

Food prices continued to rise throughout 1973, with corn prices exceeding $2 a bushel, and wheat prices remaining between $4 and $5 a bushel, despite record harvests for wheat, corn and soybeans that year. The combined effects of the massive 1972 grain purchases, increased consumption by the industrialized countries, and the maintenance of policies in the exporting countries, particularly the United States to limit production despite the changing character of world demand, had direct results in the food-importing countries of the Third World. The international food situation which emerged following the events of 1972 was not so much one of world shortage, as was widely believed, but rather one in which the price of food had risen to unprecedented levels.

Despite the unfavorable physical conditions during 1972 - bad weather over grain producing areas in the Soviet Union and Eastern Europe, the droughts in Africa and Asia, and the failure of the anchovy catch off Peru - world food production actually increased in 1972, although at a smaller rate that it had in the previous year. In addition, population continued to grow, so there were more mouths to feed. But the so-called world food shortage was not due to changes in supply, but rather to changes in demand. Increased food consumption, and the imports needed to accomplish that increased consumption in the rich countries, meant a decreased availability of food in the poor countries. The combined effects of higher food prices and the fourfold increase in the price of oil by the Organization of Petroleum Exporting Countries, and its consequent larger drain on foreign exchange, proved more than the poor food-importing countries of the Third World could cope with.

Just at the point when American food assistance was most sorely needed by these countries, the United States cut back its food aid program. The program had, in fact, been declining for a number of years. Between 1965 and 1969 and continuing through 1972, the dollar average of U. S. food assistance declined from $1.6 billion to $1 billion, the lowest level since the first year of PL 480. But because prices had risen sharply, the actual amount of food transferred declined far more dramatically than indicated. In 1957, Public Law 480-financed food exports accounted for fully 33 percent of total U. S. agricultural exports; they declined to 20 percent in 1966; 13 percent in 1972; 7 percent in 1973; and to only 4 percent in 1974. Wheat, which for many years had comprised the bulk of PL 480 shipments, had at least equalled commercial shipments until 1972. Between 1971-1972 and 1973-74, however, PL 480 wheat exports shrank from 317 million bushels to 57 million bushels. During this same period, commercial wheat exports increased from 316 million bushels to 1,092 million bushels. The countries receiving U. S. food assistance also changed during this period. By 1973, the traditional recipients among the poor, hungry countries, such as India, had been replaced by areas in which the United States had important strategic and political interests. In fiscal year 1973, nearly half of all Title I food aid went to South Vietnam and South Korea; in 1974, 69 percent went to South Vietnam and Cambodia.[4]

During 1974, however, U. S. food allocation policy gradually changed, despite the fact that food supplies remained tight, and prices high. American strategic and political objectives in Indochina had not changed either. Nonetheless, from late 1974 through 1975, amounts of food allocated under PL 480 began to expand, and larger shares of assistance were directed to genuinely needy countries. In a major policy address before the UN General Assembly in April, 1974, Secretary of State Henry Kissinger committed the United States to "a major effort to increase the quantity of food aid over the level we provided last year." And later in 1974, and again in early 1975, both President Ford and Secretary of Agriculture Earl Butz reinforced the promise to expand food aid, with redirection toward countries like India and Bangladesh.

Public Law 480 was originally formulated for the purpose of disposing of surplus food, and for developing markets for U. S. food where none existed - not, as many believed, for feeding hungry people. That it was also able to provide some hunger relief in those countries which were recipients was a positive effect of the program, but not its major objective. When PL 480 accomplished its major objective of disposing of surplus food, the consequence was higher prices. Thus the inevitable contradiction of the program - when food prices increased, and poor, food deficit countries most needed assistance, the least amount of assistance was available.

Moreover, as the commercial export market expanded, the main constituency which supported food assistance withdrew. PL 480 had been highly successful in achieving its goal of market expansion by introducing American farm products into places that would have never otherwise developed a need for them. Demand had increased substantially in the Soviet Union, China, Western Europe and Japan. Thus the agricultural community no longer had any need for PL 480. In 1974, Secretary Butz underscored this fact when he requested no funds at all for the program in the Department's budget. When questioned about this strategy, Butz is reported to have responded, "If Henry needs it, let the money come out of his budget!"[5]

During this period, the Department of State was utilizing PL 480 funds as a not so indirect means of providing military aid to the U. S. supported regimes in South Vietnam and Cambodia. For fiscal year 1974, PL 480 funds for those two countries were more than doubled from the initial request of $207 million to $499 million, an amount which accounted for more than fifty percent of American food credits extended that year. In comparison, Bangladesh, which was experiencing a severe food shortage, received only $41 million in PL 480 funding.[6]

By the summer of 1972 Henry Kissinger began to take an increased interest in the implications of food policy for U. S. foreign relations, which eventually led to an expansion of the food aid program. His first action was the commissioning of an interagency study on food and foreign policy. Also in late 1973, Kissinger, in his first major speech as the newly appointed Secretary of State, proposed the idea for a U.N.-sponsored World Food Conference, to be convened the following year.[7] It is not clear what Kissinger's real motives for the conference were at this time, considering his record of using food aid in support of military regimes in South Vietnam, Cambodia, and South Korea. Nevertheless, while Kissinger's objectives may not have been purely humanitarian, it does seem he had developed an increased awareness of the importance of internationl food issues.

Officials at the Department of Agriculture are reported not to have been in favor of the conference, because of the increased international pressure which would result on American food policies. Nevertheless, Agriculture leaders believed that by late 1974, when the conference was set to convene, predicted bumper harvests of U. S. grain would ease the supply situation, and the U. S. would be in a position to increase food assistance. Their predictions were not, however, correct.[8]

Both the Department of State and the Department of Agriculture had important stakes in developing the American position for the upcoming conference, and conflict between the agencies was plainly evident throughout the months of late 1973 and 1974. Despite the perceived pressure to increase food assistance during this period, U. S. policy remained to continue to limit it, with the bulk of aid still going to South Vietnam and Cambodia. Severe international food problems existed in South Asia and the sub-Sahara, and a worldwide fertilizer shortage aggravated the problem. Butz opposed Kissinger's effort to increase food assistance publicly, but is reported to have privately supported Kissinger, believing that the American public would not tolerate an American position which permitted starvation.

Two interagency committees shared jurisdiction over food aid allocation decisions during this period. Proposals for country by country agreements were developed by the Interagency Staff Committee chaired by USDA. But because supplies were believed too tight to allow long-range planning for allocation amounts, a second interagency committee headed by the Office of Mangement and Budget was established for the purpose of deciding on the size of agreements. State Department officials were also involved, exerting pressure for specific country allocation decisions where they regarded food aid as a necessary part of foreign policy activities. Economic policy officials, on the other hand, pressed for restrictive allocations on the basis of the continued tight supply situation, and the effects on domestic prices. Agriculture officials neither supported nor opposed either set of interests, so long as allocation increases came out of some other department's budget. When Agriculture officials began to predict bumper summer harvests, however, they began to support increased food assistance for the purpose of supporting price levels which, they feared, might begin to drop again.[9] Thus throughout the period preceding the 1974 World Food Conference, U. S. food policy was characterized by interagency conflict and a lack of a clearly defined position.

Neither the interagency study on the foreign policy implications of food policy commissioned by Kissinger, nor two subsequent studies on food aid and food reserve policy, one conducted by the Council on International Economic Policy, and the other by OMB, succeeded in clarifying U. S. policy. Kissinger had by this time developed another significant stake in U. S. food policy, as a result of OPEC's quadrupling of oil prices a few months before the Conference was scheduled to convene. He believed that increased food costs would serve to compound the worsening economic situation of the developing countries, and might move the Third World countries to put increased pressure on OPEC to lower oil prices.

The origins of Kissinger's April 15, 1974 speech before the United Nations General Assembly can be traced to the formulation of a U. S. position by Kissinger and Treasury Secretary George Shultz at the Washington Energy Conference two months earlier. The major objective of this policy was to attempt to force down the price of oil, with food policy being only of indirect concern. The strategy developed by Kissinger and Shultz was reportedly based on the idea that increased food prices would lead the developing countries to put pressure on OPEC to reduce the price of oil. Shortly before Kissinger's April speech however, a group of government officials - said to include Robert McNamara, Senator Humphrey, Peter Peterson, and members of Kissinger's staff - apparently convinced the Secretary of State to broaden the U. S. food policy position to include the overall problem of world resource scarcity. The April 15 speech committed the United States to a "major effort" to increase its levels of food assistance to the Third World, but stopped short of pledging any actual amounts.[10]

In Washington, Kissinger proposed a plan to establish a $4 billion concessional aid fund for the poorest countries suffering under the food/fuel increases. The plan was rejected by the Under Secretary's Committee, which would commit the United States to no more than $1 billion to $1.5 billion, most of which would be channeled through PL 480. The focus of U. S. food policy remained on attempting to roll back oil prices. Major opposition to the Kissinger proposal reportedly came from Treasury, which was concerned with improving the U. S. balance of payments position. Agriculture abstained from the debate, apparently believing that predicted bumper harvests would provide sufficient quantities of food for Kissinger's needs, as well as those of commercial agriculture.[11]

Kissinger apparently abandoned this strategy by early April, as grain prices began to drop, and expectations were that they would drop even further. Rather than attempting to force a confrontation, he now believed the best possibility of getting OPEC to lower oil prices would be for the U. S. to demonstrate its responsibility to the poor countries by increasing food aid. Despite resistance from economic policy officials, Kissinger promised major increases in food assistance in a speech before the Sixth Special Session of the United Nations in April.[12] And although the State Department attempted to have the Public Law 480 Budget for fiscal year 1975 increased to fulfill this pledge, the allocation figure remained at $891 million, the same amount provided in the two previous years. Anticipated declines in U. S. grain prices would, nevertheless, serve to increase the amount of aid, even though the dollar figure remained the same.[13]

This whole strategy was torpedoed by the worst summer drought in the U. S. corn belt in 20 years, and poor weather over wheat producing regions. And unlike the apparent world food shortage of 1972, when world production did not really decline, world grain production for 1974-75 declined some five percent. Food prices, which had been expected to decline, climbed at an annual rate of 13.4 percent between June and December of 1974. The administration, still reeling from the effects of devastating food price inflation, and determined not to repeat the mistake of the export embargo of 1973, naturally looked to decreasing food aid as a major means of ameliorating the price and supply situation. Congress, however, pressured by a growing coalition of interest goups committed to addressing the world food problem, continued to push for increased food aid. Since 1973, the Congress had been moving to direct increased amounts of food assistance to genuinely needy countries, and in September unanimously supported a resolution for increased food aid introduced by Senator Humphrey.

During this period, U. S. food policy decision-making was highly decentralized, with major policy debates taking place within at least three different groups. In June, 1974 the administration created the President's Committee on Food "to review and coordinate domestic and international food policy activities significantly affecting food costs and prices." (The Committee on Food took over this function from the recently abolished Cost of Living Council.) Although the Committee itself was reportedly not very active, a subcommittee, called the Food Deputies Group, headed by Gary Seevers of the CEA played an important role in deciding on and monitoring food allocation amounts for specific buyers.[14]

The second group involved was the interagency committee chaired by OMB, which continued to be responsible for decisions on PL 480 allocations and overall coordination of food assistance. Both these groups were representative of domestic economic interests and shared the objective of limiting both commercial exports and food assistance with other officials at OMB, CEA, and the Department of the Treasury. For both these groups, the major objective of U. S. food policy was the role it could play in stabilizing food prices and controlling inflation. These aims were further reinforced with the introduction of newly appointed President Gerald Ford's Whip Inflation Now (WIN) program.[15]

The third group actively involved in U. S. food policy during this period was the Department of State, which continued preparations for the American role in the upcoming food conference, under the direction of Ambassdor Edward Martin. Secretary of State Kissinger continued to support an increased American commitment to the poor countries, which directly contradicted the domestic economic objectives of the other food policy groups. Kissinger succeeded in overcoming their resistance when he persuaded President Ford to pledge increased amounts of American food assistance to needy countries in his September speech before the UN General Assembly. And although Kissinger had urged the President to commit increased amounts of food assistance, domestic economic policy officials succeeded in blocking a promise for greater volume. Thus the earlier U. S. position of not increasing the dollar amount of food aid, but increasing the total volume, was turned around.

By early fall, 1974, it had become clear that earlier Department of Agriculture predictions for bumper harvests would not come about. Both wheat and corn production fell severely, and the anticipated surplus which State, Agriculture, and economic policy officials were each counting on to meet their various policy objectives was soon over-shadowed by rising prices and forecasts of shortage. A then recently completed survey on world hunger predicted severe food problems for Africa, South Asia, Central America and the Caribbean.

During the summer, State and Agriculture had continued to debate the level of the PL 480 budget, with State pressing to increase the $900 million figure to $1.8 billion, but Agriculture resisting, and the two finally settling on a figure of $1.5 billion to $1.6 billion. When the disappointing harvest figures broke, however, it was agreed to put off a final decision. The debate continued until President Ford was scheduled to address the UN General Assembly on September 18, when he had to announce a U. S. position of food assistance. The day before the speech, Ford met with Kissinger, Agriculture Secretary Butz, OMB Director Roy Ash, and CEA Director Alan Greenspan, to arrive at a final decision.

Ford's UN speech represented a partial victory for Kissinger's foreign policy goals, with the President agreeing to pledge a "substantial" U. S. increase in food aid. Ford tentatively approved the $1.5 billion figure, which represented a compromise between Kissinger's goal of $1.8 billion, and OMB's and CEA's urging to hold the figure at $900 million. CIEP, Agriculture, and Treasury had by this time taken the middle ground on roughly the figure that was pledged, with Agriculture's major reservation being that the increase was all right, so long as the amount came out of someone else's budget.[16]

Kissinger continued to press for a U. S. commitment on increasing the amount of U. S. food assistance in preparing his own speech to the World Food Conference, scheduled to begin in November. Although Agriculture Secretary Butz was designated chairman of the U. S. delegation, Kissinger commanded considerably more prestige, both in the

Cabinet and internationally, and attempted to formulate the U. S. position independently of Butz, who was apparently neutral in the controversy. Kissinger also wished to avoid involving economic policy officials, who would surely attempt to block any attempt to pledge increased amounts of U. S. aid. But when Secretary of the Treasury William Simon finally learned of Kissinger's plan, he moved quickly to persuade President Ford to resist Kissinger's effort. Just one week prior to the date Kissinger was scheduled to deliver the official U. S. position at the Conference, Simon convened an executive committee meeting of the newly established Economic Policy Board, which was comprised of economic policy officials, and excluded representatives from either State or Agriculture. With Kissinger overseas, and unaware of these activities, Ford agreed to their recommendations, much to the chagrin of the Secretary of State.[17]

During the period of preparation for the U. S. position at the World Food Conference, still more actors had been added to the already long list of participants in the food policy arena. Despite the fact that there existed only six official slots on the American delegation, the Department of State invited twenty members of Congress to join in, although they were not even consulted on developing the U. S. stance. Formulation of an American position was further complicated by the failure between the administration and Congress to work out a general foreign assistance bill, which was a requisite to any food aid commitment before the UN. It was during this same year that Congress was attempting to enforce limits on the administration's ability to use food aid for strategic/political purposes in Southeast Asia, which had resulted in a stalemate over aid legislation in late September.[18]

In addition to changes in supply conditions since the time of Kissinger's call for the Conference, the composition of the delegation itself reflected internal disagreement among food policy officials. Kissinger and the State Department still had an interest in expanding the level of food aid for political/strategic purposes in Southeast Asia, and the Middle East. Kissinger had also apparently developed a new awareness of the enormous global dimension of world hunger, and its possible consequences for international stability. Treasury and other economic policy officials were responsible for domestic economic policy, which meant limiting food aid allocations so as to hold down U. S. food prices. And Agriculture had an interest in holding down food aid in order to serve its primary constituents, corporate agribusiness, who wanted to keep as much U. S. food as possible available for commercial purposes, particularly with the recent entries into the market by the Soviet Union and China, and increased demand in Western Europe and Japan.[19]

The failure to reach agreement prevented the administration from making a specific dollar commitment at the World Food Conference. Although Kissinger had in fact been the original proponent of the Conference, and had hoped to forge broad international agreement on all the dimensions of the world food problem, the administration's failure to pass aid legislation in effect spelled failure for the whole plan. Thus Secretary Kissinger's November speech before the World Food Conference failed to appreciably strengthen the U. S. commitment for increased food assistance that foreign policy officials had sought. The speech contained the kind of rhetoric Kissinger supported, and pledged that the United States would increase its food aid contribution, but omitted any specific allocation commitment. And despite continued efforts by members of the U. S. delegation and other observers to press Ford to make a more concrete commitment at the Conference itself, the World Food Conference adjourned without a U. S. decision on food aid for the fiscal year which was already nearly half over. Neither of the decision groups had succeeded in overcoming the opposition of the others, and no one position was adopted. U. S. food policy remained confused and without a clear direction or purpose.

It was not until February, 1975 that the administration finally announced its food aid allotment, which was set at $1.6 billion. During the period in which the decision was deliberated, certain developments occurred which led economic policy officials to soften their position on limiting food aid and, moreover, led Kissinger to press even harder for increased allocations. Economically, the Soviet Union and China decreased their purchases some 800,000 tons below the amounts originally contracted for, and prices declined.[20] In addition, Congress had in the previous year begun to move to limit the overt political use of food aid by passing legislation, as part of the 1973 Foreign Assistance Act, which prohibited the use of local currencies earned from food sales for common defense purposes. In 1974, Congress further inserted a section in that year's Foreign Assistance Act which directed the president to allocate a maximum of 30 percent of concessional food aid in fiscal year 1975 to those countries not included on the United Nation's "Most Seriously Affected" (MSA) list, unless it could be demonstrated that the food aid was for strictly humanitarian purposes.[21]

The 30/70 stipulation, known as the Humphrey amendment, threatened to place a real limitation on Kissinger's ability to use food aid for strategic/political purposes, and led to a good deal of scrambling by the administration. In January, 1975, for example, Secretary of State Kissinger personally attempted to have the United Nations reclassify South Vietnam as an MSA country in order to make it eligible for humanitarian aid, and not subject to the 30 percent limit. In yet another attempt to avoid the restriction, the administration attempted to increase the amount of food aid available for political purposes by proposing that the 30 percent limit apply to total assistance under both Titles I and II, but this was rejected by the Senate Foreign Relations Committee. Having failed in these attempts, the administration simply increased the total food aid budget to $1.6 billion, thereby making available more total aid for political purposes, while still being able to remain within the 30/70 guideline.[22]

The Development of a U. S. Position an an International Food Reserve System, 1974 - 75

United States policy opposed the creation of an international grain reserve throughout the period between the dramatic change in the world food situation in 1972, until early 1974, when certain sectors of the government began to alter their positions. Elimination of the grain reserve had been a major component of the transition from a world food system dominated by a concessional market, to one dominated by a commercial market.[23]

The dismantling of the grain reserve system was a major component of the Nixon Administration's policy of transforming U. S. food policy and, ultimately, the international food system itself. The purpose of maintaining large reserve stocks of grain had changed a number of times over the years. Originally conceived as a means of maintaining the incomes of farmers and stabilizing prices as far back as the depression, the reserve system later assumed the function of, and thereafter came to be depended upon, as a means of insuring a minimal level of food in times of shortfalls in production. Excess food played a major role in providing relief to the countries of Western Europe which had been devastated by the war, and, as the United States emerged as a world power, provided critical relief to starving people around the world in times of emergency.

But a large food reserve system still acted to stabilize prices, and the goal upon which it was originally conceived some forty years earlier was the very reason the Nixon Administration sought to abolish it. The agricultural strategy as developed by the Williams Commission and integrated into the NEP called for maximizing exports; and corporate agribusiness, which had gained major influence in Earl Butz's Department of Agriculture, strongly opposed the stable price structure which acted to limit their profits. The Department demonstrated very little concern for the consequences that eliminating the grain reserve system might have for hungry people. As Assistant Secretary of Agriculture Phil Campbell stated in a speech on July 31, 1974:

> There's no more justification for the government to store six months to a year's supply of farm goods than there would be for the government to store goods such as automobiles, dishwashers, clothes, steel, building materials, or fuel.[24]

The Nixon Administration undertook a strategy to eliminate the reserve system, beginning with the sale of American grain held in storage to the Soviet Union in 1972. The Farm Bill enacted by the administration the following year contained certain provisions designed specifically to deplete the reserve. The 1973 legislation altered the basic price support system to farmers by shifting from a loan rate system, under which the government automatically acquired farmers' crops in surplus situations; to a target price system, which was designed to allow grain corporations to purchase surplus crops at cheap rates subsidized by the Federal government.

The first sign of a new U. S. position on the international grain reserve issue was exhibited in Secretary of State Kissinger's speech before the Sixth Special Session of the UN General Assembly in April, 1974. In it, Kissinger pledged the United States "to join with other governments in a major worldwide effort to rebuild food reserves."[25] In President Gerald Ford's September, 1974 speech before the UN General Assembly, he voiced a similar promise to "negotiate, establish, and maintain an international system of food reserves."[26]

The Secretary of State continued the same theme in his speech before the World Food Conference in Rome on November 5. He characterized the "threat of famine, the fact of hunger," as the most fundamental problem facing the planet, and the convening of the WFC as recognition "that this eternal problem has now taken on unprecedented scale and urgency and that it can only be dealt with by concerted worldwide action." He spoke of the interdependence which had brought to the the world great progress, but now "threatens us with a common decline." He challenged the nations of the world to confront the crisis, and cooperate in solving it. The United States position of food abundance, he said, was a "global trust."[27]

Kissinger characterized the world food problem on two levels; first, it was necessary to cope with the emergency of famine; and second, with the long-term problem of assuring adequate supplies and standards of nutrition for the world's poor. He spoke of a growing world population, of rising expectations, and of problems of distribution and production. There was also, he stated, the problem of reserves:

> Protection against the vagaries of weather and disaster urgently requires a food reserve. Our estimate is that as much as 60 million tons over current carryover levels may be required.[28]

Kissinger offered a comprehensive program for urgent action, and cooperation by all the nations of the world on five fronts:

> Increasing the production of food exporters;
> Accelerating the production in developing countries;
> Improving means of food distribution and financing;
> Enhancing food quality; and insuring security against food emergencies.[29]

First, he called on the other major food producing countries of the world to join the United States in expanding their production, and to put an end to policies which were based on the idea that full production created undesirable surpluses and depressed markets. Kissinger declared:

> It is now abundantly clear that this is not the problem we face; there is no surplus so long as there is an unmet need. In that sense, no real surplus has ever existed.[30]

He pledged an all-out effort by the United States to expand production which, he stated, had already reached 167 million acres under grain, an increase of 23 million acres from two years previous. And in addition to steps which he called upon the developing countries to undertake to increase their own production, to improve food distribution and financing, and to improve food quality - Kissinger proposed the idea of a cooperative multilateral food reserve system, of as much as 60 million tons over present carryover levels, and including six major elements: exchange of information between countries on stocks, predicted levels of production, and planned trade requirements; agreement on the size of an emergency reserve; sharing of responsibility for maintaining reserves; agreed upon guidelines for managing national reserves; preference for cooperating countries in access to reserves; and procedures for settling disagreements and measures for dealing with noncompliance.[31]

Butz's address the following day, on November 6, stood in marked contrast to Secretary of State Kissinger's on the issue of establishing a world food reserve. Quite simply, the Secretary of Agriculture ignored the reserve issue in his address, even though global food security was slated to be a major agenda item at the Conference. The focus of Butz's speech, was on production:

> The number one responsibility of this Conference is to move the world toward a higher level of food production . . . .
>
> There are other subjects to consider, of course. There is the matter of food reserves. There is the question of emergency aid . . . . These, however, are issues that arise after food is produced - not before. We are not here to talk about what to do with less food. We are here to talk about what to do with more food.[32]

Butz devoted his address to discussing the goal of increased production, the profit incentive, and return on investment. And although he pledged support of the concept of an internationally coordinated, but nationally held food reserve system, he did so only with obvious reservation. Butz stated:

> The other subject that has come to the fore, along with food aid, is the question of food reserves. As I have already noted, the best assurance of food security is increased production. We cannot conjure a reserve out of something we don't have. To lock

> away a part of current short food supplies in order that the future might be more secure would call for less consumption this year, higher food prices, and more inflation. There are consequences that few nations would wish to entertain at the present time.
>
> . . . We will cooperate in reasonable international efforts to sustain food reserves to meet emergencies. We do not favor food reserves of a magnitude that would perpetually depress prices, destroy farmer incentives, mask the deficiencies in national production efforts, or substitute government subsidies for commercial trade.[33]

It was not until fully a year had elapsed after the World Food Conference, however, that an official U. S. position emerged on the issue of a world food reserve. On September 29, 1975, Deputy Assistant Secretary of Agriculture Richard E. Bell set forth before the International Wheat Council, an American proposal for an international reserve to consist of only 30 million tons, with responsibility for holding the reserve to be shared among the participating countries. The U. S. plan was not received very favorably, particularly by European countries who preferred a different sort of system.[34] In any event, the world food shortage which had precipitated the reserve issue had largely receded, and supplies were again increasing, and prices declining. The United States proposal for an international food reserve, which had been trumpeted loudly one year earlier, was too little and too late.

The reserve issue provides a vivid illustration of the contradictory sets of interests present in U. S. food policy after 1972. For those segments of the policy community concerned with assuring the food security of the world's poor, as represented consistently by AID, and, less consistently, by the Department of State, the maintainence of an international food reserve system was an absolute necessity. For those segments of the food policy community whose objective was a free trade market in agriculture, and high, fluctuating prices, which benefited the grain corporations, and whose interests were at this time represented by the Agriculture Department, a world food reserve was most unwanted. The larger an international grain reserve, the more it acted to dampen, and stabilize, prices, which translated into minimal profits for the grain traders. Furthermore a minimal grain reserve, or none at all, was consistent with the foreign economic policy component of the Nixon Administration's NEP, which had turned to the maximization of agricultural exports as an important means of reducing U. S. balance of payments difficulties.

Whether the Agriculture leadership under Earl Butz was honestly representative of the interests of American farmers and consumers or, as the evidence seems to indicate, of corporate agribusiness, Butz nevertheless capitalized on producer fears by appealing to concerns for their welfare. And producers did have every reason to fear the negative domestic price consequences that a large reserve could inflict. The Secretary of Agriculture warned farmers in December, 1973:

> Food reserves held by government can never be perfectly insulated from the market. . . . Farmers should not be fooled by promises that a system can be designed to protect farmers from a premature release of stocks.[35]

Agriculture Department officials attempted to appeal to still other potential sources of opposition to a government held reserve by emphasizing the cost that maintainence of such a system would entail.

The foreign policy community, however, represented a different set of interests on the issue, and had an interest in promoting the establishment of an international reserve. One reason related to the international consequences of the 1973 soybean embargo, in which the decision had failed to consider foreign policy concerns. More importantly, however, was a developing awareness on the part of State Department officials, and Henry Kissinger in particular, of the need to pay more attention to the demands of the Third World. This was a period of the opening volleys of the North/South dialogue, and of the call from the poor counties of the world for a "new international economic order." Foreign policy officials apparently recognized the need for some sort of constructive response to these demands, and that cooperation on the creation of an international food reserve was a positive, yet fairly inexpensive gesture in this regard. This policy direction continued, of course, to receive support from the development-oriented sectors of the foreign policy establishment, such as AID.

Most sources agree that Kissinger did undergo something of a transformation on the international food issue after 1973. The NSC interdepartmental study which Kissinger commissioned at the end of 1973 appears to have educated Kissinger in the importance of the relationship between agriculture and foreign policy, and an increased concern for the problem of world hunger.[36] Bearing in mind Kissinger's overall policy positions, one would have reason to suspect that this concern was based more on the possible consequences of world famine for international stability, than on humanitarian interests. The Secretary of State had come to increasingly appreciate the possible uses of food resources in foreign policy, and moved to gain further control over the area.

It has been reported that a rough consensus had been reached within the administration on the reserve issue early in 1974, having settled on a position in support of some sort of international arrangement controlled by the member states themselves, rather than by an international agency. Butz, however, continued to emphasize production as the solution to the world food shortage, and Kissinger, international cooperation. But by the Spring of 1974, the Nixon Administration was being rocked by Watergate, and a substantial segment of key economic policy officials had been replaced by new faces. In May, William Simon had become the new Secretary of the Treasury, and Kenneth Rush the new White House economic coordinator, positions previously held by the highly influential George Shultz. Peter Flanigan, who had been extremely important in the formulation and implementation of the NEP, and who had been Director of the Council on International Economic Policy, also left the administration. The CIEP under Flanigan had served as a powerful proponent of domestic economic policy objectives on food and other international economic policy issues, which often acted to counterbalance Kissinger's national security goals.[37] In June a new high-level food policy group, the President's Committee on Food, was established. Chaired by Rush, it included representatives from State, Agriculture, Treasury, OMB, and the CIA. In turn, the food committee of the Cost of Living Council was phased out. Any coherence which might have survived this chaos was probably finally lost with Nixon's resignation from the Presidency in August. Newly appointed President Ford created his own group to coordinate economic policy a month later, the Economic Policy Board (EPB), and this in turn superseded the President's Committee on Food.[38]

The fact that the world food reserve issue was before a political body, the United Nations, rather than an economically or agriculturally-oriented international body was believed at the time to give the State Department the upper hand over Agriculture, and allowed State to be put in charge of the American delegation. It was also widely believed that despite his opposition to such a formula, that Agriculture Secretary

Butz had conceded to some form of government-held reserve system. Nevertheless, the upcoming jockeying between State and Agriculture over an issue which both agencies had come to perceive as being vital to the set of interests each represented appeared inevitable. Daniel Balz called the State and Agriculture Departments the Hatfields and McCoys of the government. He further commented, in a 1975 article:

> The policy fight in this country has pitted Secretary of State Henry A. Kissinger and his desire to push the United States into a leading role in solving international food problems against Secretary of Agriculture Earl L. Butz and his resistance to any policy that threatens the commercial export of American farm products or leads to unrest in the farm community.[39]

Kissinger's opening address to the World Food Conference seemed to indicate that State had won out over Agriculture on the international reserve issue. Although Secretary of Agriculture Butz's speech clearly indicated his lack of enthusiasm for a reserve, he did acknowledge that official U. S. policy supported the creation of "an internationally coordinated but national held" system.[40]

Although the grain reserve system had long been under the Agriculture Department's jurisdiction, Kissinger recognized that an important element in wresting control of the issue away from Agriculture depended on defining the international food reserve question as a foreign policy issue. The Secretary of State's first maneuver was to present the U. S. position in Rome as one which strongly supported a reserve system, despite the considerable interagency conflict which existed within the American delegation. Second, Kissinger moved to claim responsibility for coordinating follow-up measures by securing President Ford's approval to create a new interagency committee for that purpose under the jurisdiction of the National Security Council, a move which was strongly opposed by both the Department of Agriculture and economic policy officials. The Committee was called the International Food Review Group (IFRG), and was chaired by Kissinger himself. Although the formal group met only rarely, an IFRG working group headed by Kissinger's Assistant Secretary for Economic Affairs Thomas Enders met frequently. These forums provided Enders with the opportunity to push State's advocacy of a meaningful reserve system before economic policy officials resistant on the issue, which included representatives from Treasury, OMB, CEA, CIEP, and the Office of the Special Trade Representative.[41]

Basically, State sought establishment of a food reserve system that contained large enough supplies to provide maintenance of market stabilization and food security when necessary, and guidelines for the management of such reserves which were sufficiently specific and binding on conditions under which stocks would be added to or released. Agriculture's position lay at the opposite end of the spectrum, opposing government held stocks, and any sort of domestic supply management which interfered with a free-market situation in agriculture. The goals of price stabilization and food security sought by the foreign policy sector were exactly what agricultural interests wished to avoid.

Economic policy officials, in the meantime, had begun to move closer to Agriculture's position and, consequently, further from State's. The Council of Economic Advisors which had nominally supported the establishment of an international reserve system prior to the World Food Conference, was by February 1975 less enthusiastic over the conclusion of such an agreement. The position held by economic policy officials could not be classified as simply for or against an international reserve. On the one hand, such a system, if fully implemented at a meaningful level, would serve to stabilize prices and therefore have a positive effect on domestic food price inflation, which was a major imperative of the administration. On the other hand, such a system would also have the appearance of being a major concession to the developing countries' demand for international commodity arrangements to regulate prices, an important component of the New International Economic Order which economic policy officials opposed.

CEA officials accordingly proposed a compromise measure which would have established a separate, limited food reserve for the developing countries. This idea was opposed by State, which sought the international reserve as an instrument for stabilizing agricultural relations with Europe and Japan, which had suffered severely as a result of the soybean embargo; and with the Soviets and Chinese, for which agricultural trade was a principal underpinning of more favorable relations in other areas. Kissinger had also apparently developed a concern for the potential international political unrest which a world food shortage would either cause, or at least exacerbate. The CEA strategy was also opposed by Agriculture, however, which viewed the plan as an unwanted linkage between the reserve system and food aid, which Agricultural wished to maintain as small as possible. However logical the CEA proposal may have been, it was doomed to bureaucratic failure.

Part of the State Department's inability to have much bureaucratic success in advancing its position versus that of Agriculture and economic policy officials was the fact that within the Department, the fight was carried almost solely by Kissinger. Food reserve policy had belonged fairly exclusively within Agriculture's domain up until this time, and thus there existed very little organizational interest or experience on the issue beyond Kissinger and Enders. Following the negative impact on U. S. foreign relations with Japan and Europe which had resulted from the soybean embargo, Kissinger had commissioned a number of studies on the relationship between food and foreign policy. Partially as a result of those studies, Kissinger had developed a heightened sense of awareness on the importance of food policy, and particularly on the significance which food aid and reserves had for the stability of the Third World. The Secretary of State operated more from his personal power base and as the President's national security advisor, however, and made minimal use of the State Department bureaucracy. This undoubtedly contributed to the fact that Kissinger's concern for U. S. food policy and the relationship to international stability apparently had little organizational support. According to Balz:

> The Agriculture Department has been successful in the interagency debate partly because it has put up a more united front than has State . . . .
>
> . . . (Kissinger) was almost alone in trying to boost U. S. food aid commitments before the World Food Conference, and he has been the senior official within the Ford Administration most concerned about building an international reserve system.[42]

Kissinger also believed that an international food reserve system would serve to counterbalance the instability that unpredictable Soviet grain buying habits brought to the world food system. Still another factor may have been the role a U. S. position of leadership on the international food issue could have played in helping to restore American prestige in the aftermath of Vietnam. But Kissinger's policy style was highly personal, and while he was very effective on many foreign policy issues, the reserve question was, despite his attempts to frame it otherwise, more rightly an economic issue. Kissinger had neither the expertise nor organizational support to gain jurisdiction over the issue.

Thus the Kissinger-Enders International Food Review Group failed to accomplish its mission of negotiating a compromise on the international reserve issue. As a consequence, the Department of State lost the ball, and the issue was transformed from a foreign policy problem into an economic policy problem. On June 11, 1975 the President's Economic Policy Board directed the State Department to prepare a new options paper.[43]

The Economic Policy Board has no formal role in developing a policy on the reserve issue, but was responsible for coordinating overall economic policy for the administration. The EPB became involved when OMB Director James T. Lynn apparently lost patience with IFRG's ability to coordinate a decision, and finally agreed to intervene on the matter, something which the OMB staff had reportedly pushed for months.[44]

By mid July an official U. S. position was finally agreed upon. As announced by Kissinger before the Seventh Special Session of the UN General Assembly on September 1, and subsequently proposed to the International Wheat Council on September 29, the U. S. proposal for a reserve system was for only 30 million tons, half of what Kissinger had announced as necessary in his WFC speech almost a year earlier.[45]

Ironically, it was not until the point where State officials lost control of the issue to economic policy officials that an internal consensus was reached on the international grain reserve question. The IFRG which Kissinger had initially established to gain control of the issue and cast it as a foreign policy problem proved ineffectual in negotiating a compromise with Agriculture and economic policy officials, who probably distrusted the group as being too biased in favor of State's position. Despite the bureaucratic control of the issue State was successful in exerting, based on the formation of the IFRG as the approved forum for settling the issue, and Kissinger's personal power, the international food reserve issue was too important to agriculture and economic policy officials to relinquish. A U. S. policy position simply had too many implications for important agricultural and economic policy interests to permit the issue to be controlled within State's jurisdiction.

Chapter Notes

The Use of Food Resources in U. S. Foreign Policy Toward the Third World, 1974 - 75

1. Leslie Gelb and Anthony Lake, "Washington Dateline: Less Food More Politics," Foreign Policy 17 (Winter 1974-75), p. 177.

2. See previous section on 1972 Soviet grain deal for discussion of this.

3. Gelb and Lake, "Less Food," p. 178.

4. I. M. Destler, Making Foreign Economic Policy (Washington, D. C.: Brookings Institution, 1980), p. 66.

5. Ibid., p. 68.

6. NACLA, "U. S. Grain Arsenal," Latin America and Empire Report 9 (October, 1975), p. 13.

7. Destler, Foreign Economic Policy, pp. 69-70. The idea for an international forum for the purpose of addressing the problem of food shortages in the poor countries was first advanced by Senator Humphrey during Kissinger's nomination hearings. Destler believes that Kissinger had by this time developed a "secondary interest" in humanitarian food aid. For a similar account of the development of the American position at the World Food Conference, see Thomas William Weber, Agricultural Exports and Decision-Making in American Foreign Policy (Unpublished Ph.D. dissertation, University of Virginia, May, 1977), pp. 290-299.

8. Destler, Foreign Economic Policy, p. 70.

9. Ibid., p. 71.

10. Gelb and Lake, "Less Food," p. 180.

11. Ibid., p. 181.

12. Department of State Bulletin, 70 (May 6, 1974), p. 480.

13. Destler, Foreign Economic Policy, p. 72.

14. Ibid., p. 73.

15. Ibid., pp. 73-74.

16. Gelb and Lake, "Less Food," p. 184.

17. Destler, Foreign Economic Policy, p. 75. For another account of these events, see Daniel Balz, "Agriculture Report: Politics of Food Aid," National Journal, p. 1762.

18. Gelb and Lake, "Less Food," p. 185.

19. Martin Kriesberg, "Food Aid and Foreign Policy," in Ross B. Talbott (ed.), The World Food Problem and U. S. Food Politics and Policies: 1972-1976 (Ames, Iowa: Iowa State University Press, 1977), pp. 162-163.

20. Destler, Foreign Economic Policy, p. 76. Actual food aid expenditures for fiscal year 1975 reached only $1.2 billion.

21. U. S. House of Representatives, Committee on International Relations, Use of Food Resources for Diplomatic Purposes (Washington, D. C.: U. S. Government Printing Office, 1977), p. 25.

22. While Kissinger in effect circumvented the 30 percent limit by increasing the total amount of food aid in order to make available more for political purposes, it nevertheless also acted to increase the aid total for humanitarian purposes as well, thereby accomplishing at least one of Humphrey's objectives. See NACLA, "U. S. Grain," p. 14; and Susan DeMarco and Susan Sechler, The Fields Have Turned Brown, (Washington, D. C.: The Agribusiness Accountability Project, 1975), pp. 41-42.

The Development of a U. S. Position on an International Food Reserve System, 1974 - 75

23. The world's food reserve capacity is held in two principal sources: the carryover reserve stocks of grain held by the major exporting countries, the most important of which is the United States. In 1961, these reserves represented the equivalent of 222 million tons of grain, or about 95 days of the world's consumption. In 1974, however, these reserves repesented only 26 days of the world's food needs. See, for example, Lester R . Brown, By Bread Alone (New York: Praeger, 1975), Chapter 1.

24. DeMarco and Sechler, The Fields Have Turned Brown, p. 23.

25. "The Challenge of Interdependence," Department of State Bulletin 70 (May 6, 1974), p. 480.

26. "Address to the 29th Session of the General Assembly of the United Nations, September 18, 1974," cited in Deslter, Foreign Econommic Policy, p. 89.

27. Address by Henry Kissinger before the World Food Conference, "The Global Community and the Struggle Against Famine," reprinted in U. S. Senate, Committe on Agriculture and Forestry, Subcommittee on Foreign Agricultural Policy, Hunger and Diplomacy: A Perspective on the U. S. Role at the World Food Conference (Washington, D. C.: U. S. Government Printing Office, 1975), p. 105.

28. Ibid., p. 107.

29. Ibid., p. 108.

30. Ibid.

31. Ibid., pp. 108-114.

32. Address by Agriculture Secretary Earl Butz before the World Food Conference, "Let's Become Hunger Fighters," reprinted in U. S. Senate, Hunger and Diplomacy, p. 116.

33. Ibid., pp. 117-118.

34. Destler, Foreign Economic Policy, pp. 89-90.

35. Cited in Stephen S. Rosenfeld, "The Politics of Food," Foreign Policy, (Spring 1974), p. 24.

36. Ibid., p. 27.

37. Destler, Foreign Economic Policy, p. 93.

38. Ibid., pp. 93-94.

39. Daniel Balz, "State-Agriculture Feud Delays Grain Reserve System," National Journal, June 28, 1975, p. 951.

40. See address by Butz, reprinted in U. S. Senate, Hunger and Diplomacy, p. 118. Butz seemed to underline his personal opposition to the reserve even while including mention of the official U. S. position. Apparently unable to say the words himself, he inserted a quote on the matter by President Ford supporting creation of a reserve, but sandwiched it between strongly negative statements on possible consequences of a reserve system.

41. Destler, Foreign Economic Policy, p. 97.

42. Balz, "State-Agriculture Feud," p. 959.

43. Destler, Foreign Economic Policy, p. 101. Also, by this time, CEA's agricultural specialist Gary Seevers had been replaced by Paul MacAvoy, who also succeeded Seevers as chairman of the Food Deputies Group. MacAvoy ordered a new study of the reserve issue, which concluded that establishment of an international food reserve system would have a positive impact on the global food system.

44. Balz, "State-Agriculture Feud," p. 953.

45. Destler, Foreign Economic Policy, p. 101.

CHAPTER 7

THE SOVIET GRAIN EMBARGO, JANUARY 4, 1980 - APRIL 24, 1981

The grain embargo directed against the Soviet Union from January 4, 1980, through April 24, 1981, actually offers a glimpse into two decisions. The first decision involves the Carter Administration's imposition of the embargo, while the second involves the Reagan Administration's removal of the embargo almost sixteen months later. The decisions are particularly interesting because the first was made for almost completely foreign policy purposes, while the second was determined largely by domestic reasons.

The Soviet Grain Embargo, January 4, 1980

On January 4, 1980 President Jimmy Carter, denouncing the recent Soviet military intervention in Afghanistan as an "extremely serious threat to the peace," announced a set of U. S. actions against the Soviet Union, including a cut-off of sales of high technology goods, a curtailment of Soviet fishing privileges in American waters, and the withdrawal of seventeen million tons of American grain that had been on order by the Soviet Union. Another eight million tons of grain, which the United States was required to sell the Soviets under the terms of the 1975 grain agreement, and most of which had already been delivered, were not to be affected.[1]

American officials did not really expect that any of these measures would act to force the Soviet Union to suddenly withdraw its forces from Afghanistan. Rather, the steps were designed to put the Soviet leadership on notice that such military activities would not be completely ignored by the United States, that business between the two countries would not continue as usual. In announcing the action, President Carter stressed that he had fully consulted with congressional leaders, as well as with America's European allies prior to the decision, and had been assured of support by both groups. Carter had been repeatedly criticized on previous foreign policy decisions for failing to do either.[2]

The language of Carter's announcement indicated that the measures were in no way intended to pose an ultimatum to the Soviets, and left open the possibility for future improvement of relations between the two countries. State Department officials expressed that the American actions were meant to communicate that "the world simply cannot stand by and permit the Soviet Union to commit this act with impunity." One senior official expressed the Department's position:

> We've seen things in the past that dissatisified us, in Southeast Asia, with the Cubans, but the Soviets crossed a new threshold in Afghanistan, and we have entered a new threshold in our actions, which amounts to a real bite on the leadershihp and people.

The Soviet people and leaders, the spokesman contended, had grown accustomed to benefitting economically from U. S. and other Western countries' agricultural products and industrial technology.

> I think, therefore, that the kind of decisions we have made to restrict trade will make them aware that there is a serious price and make them think twice about what they do in Afghanistan and the rest of the world.[3]

As a result of consultations with other major grain-exporting countries, President Carter stated that he believed the United States would receive the cooperation it would need in order for the embargo to have its intended impact: "I am confident that they will not replace these quantities of grain by additional shipments on their part to the Soviet Union."[4] Secretary of Agriculture Bob Bergland stated at a press conference on January 5 that the United States had invited Canada, Australia, Argentina and the Commission of the European Communities to send representatives to attend an emergency conference on the matter in Washington. Bergland told reporters that he had calculated that an additional three million tons of U. S. grain might still be obtained by the Soviets, in addition to the eight million ton obligation, "through transfers that might be impossible to monitor."[5]

In announcing the Soviet grain embargo, administration officials went out of their way to stress that there was unanimous support among the President's advisers on the steps taken, including Secretary of State Cyrus Vance and National Security Adviser Zbigniew Brzezinski, who had a record of disagreeing on courses of action in foreign policy. The Secretary of State had been inclined to take a much more conciliatory position toward the Soviets, and believed it was more important to preserve detente than Brzezinski did. However, great care was taken to give the impression that the two officials were in complete agreement on the decision to impose the grain embargo, "because of the nature of Soviet aggression." Even Mr. Brzezinski went out of his way to leave the door open to restoring friendlier relations with the Soviets, characterizing the grain cut-off as a temporary measure. At his press conference he stated:

> This is not a permanent suspension. The President did not want to abrogate the bilateral purchase agreement with the Soviet Union. We hope the Soviets will come to their senses soon and we can resume our normal agricultural and economic trade with them.[6]

At least one observer believed that Secretary of State Vance and other State Department officials had been successful in building into the U. S. action the restraint that was designed to keep alive better relations with the Soviets in the future. In delivering the announcement of the embargo, the President's speech was careful not to issue any sort of ultimatums to the Soviets for withdrawal of their troops, or to threaten a complete severing of relations, or even the possibility of a total trade embargo. Although American reaction to the invasion of Afghanistan undoubtedly put a freeze in U. S.-Soviet relations, the actions were designed to not necessarily be permanent, and clearly left the way open for a reconciliation. President Carter's position seemed to be based on the belief that the Soviet actions in Afghanistan could not realistically be reversed but, nevertheless, it was important to notify the Russians that relations with the United States would be considerably strained.

Furthermore, the decision to withhold delivery of some seventeen millions tons of grain to the Soviet Union was not really expected to have an immediate impact on their food supply. Neither was the measure expected to immediately affect their ability to supply grain to their Communist bloc clients, including Afghanistan.. All of the U. S. grain which the Soviets had expected to import, which would have amounted to a total of 25 million tons if the embargo decision had not been made, was expected to go to feed expanding livestock herds in the Soviet

Union. U. S. experts believed the Soviets had more than enough wheat to feed their people, some livestock, and still export about 500,000 tons, the same amount as in the previous year, despite a 30 percent drop in the fall grain harvest.

Furthermore, American officials were aware that there were a number of alternatives to which the Russians could resort in order to offset the loss of the additional American grain. They could speed up the slaughter of livestock, as they had done in previous shortage years. Or they could alter the type of feed they were using, which would mean purchasing more expensive soybeans, probably from Brazil. Or they could increase their poultry flocks, which require far less grain than cattle.[7] Still another variable was the size of the Soviet Union's own grain reserves, which were believed to be quite healthy as a result of a favorable harvest in 1978, but were kept a well-guarded secret.

President Carter was obviously concerned over the impact that the grain embargo decision might have on his political standing at home. The presidential election year was just a few days old, and the administration had obviously taken into consideration the possible loss of farm-state votes in reaching its decision. Nevertheless, foreign policy officials apparently agreed that some form of action had to be taken to demonstrate to the Soviets that the United States was not completely without sources of leverage in a situation which did not pose a direct security threat to American interests. Although some observers argued that the taking of Afghanistan represented as stepping stone on the way to Middle East oilfields, no one expected the United States to risk a direct confrontation over a remote area which was already within the Soviet sphere of influence. The United States had emerged from the recent Iranian revolution looking rather impotent - the American hostages were still being held by militant students - and some experts felt this show of weakness had in fact prompted the Soviets to invade Afghanistan. Thus, some foreign policy experts believed it was imperative for the United States to take some form of action, even if the form chosen was not popular with certain groups at home. Apparent American impotence in the Iranian situation created the need for President Carter to demonstrate to the American public a willingness to act as well. The President's domestic political advisers therefore believed, for reasons similar to those of his foreign policy advisers, that some form of action was necessary. In his address to the nation, President Carter emphasized what he portrayed to be the very serious nature of the Soviet action:

> This invasion is an extremely serious threat to peace - because of the threat of further Soviet expansion into neighboring countries in Southwest Asia, and also because such an aggressive military policy is unsettling to other peoples throughout the world.
>
> This is a callous violation of international law and the United Nations Charter.
>
> It is a deliberate effort of a powerful atheistic government to subjugate an independent Islamic people.
>
> We must recognize the strategic importance of Afghanistan to stability and peace.[8]

At the same time, however, the President moved to firmly reassure American farmers that the government would take whatever steps were necessary to ensure that prices of agricultural commodities would not fall. Thus he declared later in his address:

> I am determined to minimize any adverse impact on the American farmer from this action. The undelivered grain will be removed from the market through storage and price support programs and through purchases at market prices. We will also increase amounts of grain devoted to the alleviation of hunger in poor

countries and will have a massive increase of the use of grain for gasahol production here at home.[9]

Despite the steps announced by President Carter to soften the immediate impact of the Soviet grain embargo on American farmers, which were expected to cost U. S. taxpayers some 2.5 to 3 billion dollars over a 21 month period, the withdrawal of the grain shipments was expected to hurt the overall U. S. economy, as well as farmers.[10] Even though the government had promised to purchase the grain which would have been exported to the Soviets, that grain would still hang over the market and depress prices. And unless alternate buyers could be found for the grain, some $2 billion, which the grain would have earned overseas, would be recorded on the minus side of the overall U. S. balance of payments, which had registered a $24 billion deficit the previous year.[11] Since 1972, agricultural exports had provided the major source of earnings in the nation's trade balance. The grain embargo would, in any case, present a major budgetary expense for the Federal government during a period which demanded severe fiscal restraint.

Many observers predicted that the long-term effects of the embargo on the U. S. economy could also be quite serious. If the Russians slaughtered large numbers of livestock due to a lack of feed grain, future demand for U. S. grain would be severely reduced. Other grain exporting states might try to fill the vacuum left by the U. S. action, and increase their exports to the Soviet Union in future years. The United States' reputation as a commercially reliable supplier of agricultural commodities would surely be brought into question. Since moving to a commercially dominated food export system in 1972, very few observers would have predicted that the United States would choose to exercise food policy for political purposes on the scale involved in the Soviet embargo.

Officials within the administration disagreed over how seriously the embargo would hurt the Soviets. It could not be determined how much of the deficit would be made up by other countries; nor was it known how full Soviet grain elevators were after their record harvest in 1978. Some officials estimated that other grain exporting countries could supply the Soviets with an additional seven million metric tons of grain. This amount could be affected by American success in persuading other countries to respect the embargo. Another variable would be how the Soviets chose to deal with the situation internally.

In any event, the U. S. farm community feared the real impact of the embargo would be felt in the U. S. in both the short and the long term. Farm organizations believed the action would lead other countries to question America's reliability as an exporter, as happened after the 1973 soybeam embargo, when Brazil emerged to permanently seize a major share of the international soybean market from the United States. John Block, Illinois Agriculture Director, and later President Reagan's Secretary of Agriculture, stated:

> The Soviet Union is never going to count on us again, and I don't blame them. It will damage our potential sales to the Soviet Union for years to come.

Allan Grant, president of the American Farm Bureau Federation was typically irate at the administration. He commented:

> Mr. Carter took aim at the Russians with a double-barreled shotgun, pulled the trigger and hit the U. S. farmer.[12]

Secretary of Agriculture Bob Bergland moved quickly to reassure American farmers that they would not suffer any loss of income due to the U. S. grain embargo against the Soviet Union. He said the administration would take whatever steps were necessary to ensure that grain prices did not fall. Many U. S. farm organizations cautiously supported the use of the embargo in response to the Soviet Union's invasion of Afghanistan, but were nevertheless skeptical of the administration's ability to prevent grain prices from declining, no matter what measures it took.

At a news conference on January 5, Agriculture Secretary Bergland insisted that the grain embargo was not a case of using food as a diplomatic weapon, something which President Carter had previously promised American farmers he would never do. This idea was based on his interpretation that food only becomes a weapon when it is used to cause hunger in a targeted country, and it is different when the grain is denied for purposes of feeding cattle. Bergland stated:

> Food becomes a weapon when you deny it to the people of a hungry nation. This is not the same. The Soviet Union has invaded a free, sovereign country and we are not going to support them in this with our grain.[13]

Although this line of reasoning appeared to be rather thin, both Secretary Bergland and President Carter had previously gone on record as promising that the administration would never employ food as a political weapon. Mr. Bergland, in fact, at one time had likened the use of food as a diplomatic weapon to "shooting oneself in the foot." President Carter came under immediate attack from his Republican challengers, as well as Democratic contenders for the presidency, over the issue of the Soviet grain embargo. The Iowa primary caucauses were scheduled to be held for both political parties less than three weeks after the grain embargo was announced, and President Carter's opponents seized the opportunity to attempt to sway votes. Many farmers had supported

Mr. Carter in 1976, and he had promised at the time to never employ grain embargoes.[14]

Secretary of Agriculture Bergland spent the period between the embargo decision and the Iowa caucauses trying to shore up farmers' votes by appealing to their patriotism, and by counting on their traditional support for taking a hard line against the Soviet Union on strategic matters. Bergland kept repeating to Iowa farmers: "We're all being tested by the Russians to see what we're made of, and they are finding Jimmy Carter is made of steel."[15]

Farmers nevertheless appeared highly skeptical of the measures the administration was planning to protect their incomes from the effects of the embargo - buying up the undelivered sales contracts held by the major grain traders, increasing the grain reserve, and a plan to use surplus grain to produce gasahol. Many farmers were inclined to believe that most of the undelivered grain was likely to end up in storage bins, and no matter who owned it, would act to depress prices. Farmers also believed that the embargo would serve in the long term to reduce future Russian dependence on U. S. grain, even if it was subsequently lifted.

The decision to embargo grain and technology to the Soviet Union represented a major reversal by the Carter Administration on the belief in the usefulness of trade as a political weapon. At the time of the grain embargo, administration officials admitted that the economic warfare strategy adopted was strongly contested within their ranks, which previously held the position that such actions were likely to cost the United States more economically than they gain politically.

The prinicpal exponent of a tough economic warfare strategy against the Soviets within the Carter Administration was reported to be assistant to the president for national security affairs Zbigniew Brzezinski. Brzezinski had apparently supported the withholding of trade with the Soviet Union as a means of causing them economic problems and, he argued, thereby weakening them as a stretegic threat. Mr. Brzezinski had been supported in this position by former Secretary of Energy, and then Secretary of Defense James R. Schlesinger. Defense Department officials reportedly supported controlling any goods which might possibly be used for strategic purposes.[16]

On previous occasions Mr. Brzezinski's position had had mixed results. In 1978 he attempted but failed to block a sale to the Soviet Union of oil drilling technology during a highly publicized trial of Jewish dissidents. He was successful, however, in blocking a smaller computer sale to the Soviets, and was able to place oil drilling equipment on the list of exports which required official licensing. In addition, according to administration officials, Brzezinski, with support from the President's chief of staff Hamilton Jordan, succeeded in preventing the nomination of Stanley J. Marcus to become Deputy Assistant Secretary for Trade Regulation in the newly organized Commerce Department. Mr. Marcus had opposed Brzezinski on blocking a sale of technology to the Soviets by Dresser Industries.[17] The National Security Adviser and the Commerce Department had consistently come down on opposing sides on the whole issue of using trade for political purposes, as would have been expected.

U. S. foreign policy officials clearly did not believe that the series of economic measures taken against the Soviet Union would induce them to alter their course of action. Rather, U. S. policy was designated to demonstrate that the United States would not stand idly by in the face of Soviet aggression. Economic policy was viewed as a middle-ground between military response, which of course was not really considered over the Afghanistan situation, and doing nothing. According to testimony by Special Adviser to the Secretary of State on Soviet Affairs Marshall D. Shulman:

> . . . it is absolutely necessary to indicate to the Soviet Union that the measures that we are now taking indicate the seriousness with which we take this action, that they are not simply going to be forgotten in a month or two, that they are going to be sustained.
>
> It would be, in our judgment, a mistake to allow the Soviet Union to believe that this will blow over and be forgotten.[18]

State Department officials, nevertheless, apparently had a rather optimistic view of the grain embargo in terms of its likely impact on the Soviet Union. Mr. Shulman testified that he believed the Soviets would be able to make up a maximum of only four to five million tons of grain from other sources, out of the 17 million which the United States was denying them. Other sources believed at the time that the Soviets would be able to make up most of the shortfall, although at higher financial cost. Mr. Shulman stated his belief that the grain cut-off would "hurt them quite a bit," not only in terms of their food supply, but in overall economic terms as well. He concluded:

> The effect is serious on them, across-the-board in their economic planning. It is not only going to require them to adjust their whole 5-year plan for the agricultural and the industrial imports, but it is going to be that their military expenditures are going to be a good deal higher even than they had projected.
>
> The effect on their economic planning will be very substantial. That is not something that they will take very lightly because it comes on top of economic difficulties that they have been struggling with, even before this.[19]

Secretary of Agriculture Bob Bergland, on the other hand, had calculated the cost to the Soviet Union from the embargo in more modest terms two days earlier:

> While we have calculated the impact of the suspension on Americans to be relatively small, we expect that the reverse will be true for the Soviet Union.
>
> Without 17 of the 25 million tons of grain that they had counted on, Soviet policymakers will be unable to sustain their long-term effort to improve their peoples' diets by increasing the amount of available meat. There will be substantially less grain to feed livestock - 5 to 11 percent less as a result of this suspension.[20]

In testimony before the Senate Committee on Agriculture, Nutrition and Forestry, shortly following imposition of the grain embargo against the Soviet Union, Agriculture Secretary Bergland supported the position that the 1980 embargo was taken "for reasons of urgent national security and foreign policy considerations." It was, therefore, different from the suspensions of agricultural exports in 1973 and 1975, which were enacted under the same authority - the Export Administration Act - but were taken because of short supply conditions for the affected commodities. In addition, in neither of those previous actions did the government move to mitigate potential economic losses to the American farmer, as the Carter Administration had done. Bergland expressed the administration position, that although some sacrifice would be necessary by all Americans, including of course farmers, the Agriculture Department was, at least officially, fully behind the embargo. Bergland stated:

> During times of national crisis, the American people have always been willing to make whatever sacrifices are necessary to meet the challenges which such events pose. This is a time when some sacrifice will be called for, but I am firmly convinced the actions we have taken will reduce the amount of sacrifice to much lower levels than one might expect.

> Mr. Chairman, it is my firm conviction that this action, together with other actions announced by the President, will serve notice on the Soviet Union that naked aggression against independent, nonalligned countries will not be tolerated by the United States. I strongly believe that by taking these steps, we will be accomplishing a great deal without nearly as much sacrifice on the part of the American people as other measures would have entailed.[21]

Although great care was taken by the Carter Administration to give the impression that the Agriculture Department supported the embargo, and played an equal role with the Department of State in recommending the decision, the evidence seems to indicate otherwise. According to Secretary Bergland's testimony, the recommendation to establish the grain embargo was made virtually simultaneously by the Departments of Agriculture and State.[22]

Secretary of Agriculture Bergland, however, was apparently not contacted by the President until the invasion had actually taken place. He did not appear to be aware, at that time, of any ongoing discussions within the administration over choosing a course of U. S. action. On the afternoon of the announcement of the embargo, in response to inquiries surrounding rumors of an impending embargo, the Agriculture Department was assuring farmers there would not be one, although the Secretary by this time knew there would.[23] It appears certain that the Department of Agriculture played little, if any role in the decision to impose the grain embargo. Agriculture Secretary Bergland was called in, however, after the decision was made, to help shore up the inevitable opposition by American farmers, and to provide additional credibility to the decision.

The Reagan Administration Decision to Lift the Grain Embargo

Although Ronald Reagan chose to make the grain embargo a major issue in the 1980 Presidential campaign, promising to lift it if he were elected, a few weeks after taking office the new president was reportedly reconsidering his position. Again, the forces within the administration which supported the embargo against the Soviet Union were represented by the new Secretary of State Alexander Haig, who argued that the time wasn't right for lifting it. President Reagan was also committed to a tougher foreign policy stance, particularly vis a vis the Soviets, and ending the embargo might transmit the wrong signal at a sensitive time in the new administration.

In a Cabinet meeting on February 4, 1981, Secretary of State Haig also argued that the grain embargo was, in fact, fulfilling its objective of hurting the Russians. Haig believed that the Soviets were indeed facing severe grain shortages, and were having to use scarce hard currency to purchase grain from Argentina and other countries at premium prices. Although President Reagan had said at a press conference a week earlier that he always thought the embargo was "more of a gesture than something real," an administration source reported that the President was "very impressed" with Haig's argument, and at the very least agreed that the time was not right for lifting the embargo. President Reagan had, after all, stated at the same news conference that he believed the Soviets were "bent on world domination," and considered it "moral to lie, cheat, and commit crimes" toward this end.[24]

Secretary of Agriculture John Block, like Reagan long on record as vigorously opposing the embargo, made the case for lifting the embargo at the Cabinet meeting. Block called the embargo "the most ridiculous thing I ever heard of," and stated his belief that if the Soviets had to spend their money on grain somewhere, it should be in the United States. Block had succeeded in having the issue considered by the full Cabinet, which precluded consideration of lifting the embargo solely on foreign policy or national security grounds.[25] The Agriculture Secretary's position was weakened, however, by reduced U. S. grain supplies in 1981, thereby creating a situation where export surplus was not an issue. In fact, U. S. agricultural exports had continued at record levels all through 1980, despite the Soviet embargo. Farmers maintained the position they had taken in January, 1980 when the embargo was initially imposed: they would support it, as long as they were not unduly hurt financially. And two additional factors had emerged since Reagan made his campaign pledge to end the embargo. First, there was increasing evidence that the embargo was indeed hurting the Soviets. Soviet President Leonid Brezhnev had publicly admitted that serious food shortages existed, and had assigned the highest priority to "improvement of the food supply." The new Republican Chairman of the Senate Foreign Relations Committee Charles Percy had returned from a tour of the Soviet Union with what he believed was convincing evidence that the embargo was hurting the Russians. "They talk of it as loosening belts," he stated, "but what they mean is that they can't feed their own people."[26] The Soviets reportedly had already payed out more than $1 billion of their scarce hard currency for grain from other exporting countries. And second, U. S. - Soviet tension had once again increased, this time over Soviet threats to Poland.

In a press conference the day after the February 4 Cabinet meeting, the White House reported that President Reagan had not yet decided whether to lift the Soviet grain embargo. Anti-embargo forces in the Senate, led by Senator Robert Dole of Kansas, a strong opponent of the embargo, continued to press for its removal, and were promised by Reagan that he would not make a decision on the embargo until a scheduled meeting with farm state senators on February 17.[27]

On April 24, 1981, President Reagan lifted the embargo on grain sales to the Soviet Union, voicing assurances that neither the Soviets, nor others around the world would interpret his action as a weakening of the U. S. position toward Soviet aggression. The removal of the grain embargo was the first major policy decision made by the Reagan Administration involving the Soviet Union. Reagan had promised to lift the embargo during the campaign, but had also sought to establish a tough, consistent foreign policy stance toward the Soviets, something many observers believed the Carter Administration had lacked. Thus speculation had emerged that as President, he might change his mind.

The grain embargo decision seemed highly inconsistent with increasing anti-Soviet rhetoric by the administration, including Reagan's January 29 statement charging that Soviet leaders lie, cheat, and commit crimes to achieve their goals, and the depiction of the increasing Soviet threat in the Persian Gulf region just a few days earlier. The administration offered as partial justification for the action the lessening of tensions in Poland, where a Soviet invasion had been feared. Reagan also stated, as he often had before, his belief that the embargo had not hurt the Soviets, and that it had placed an unfair burden on American farmers. The President said that lifting the embargo had been under constant review since he had assumed office, and that he believed the action was not contradictory or inconsistent with overall policy toward the Soviet Union. He argued that it would not send the wrong signal to the Russians, stating:

> In the first few weeks of my presidency, I decided that an immediate lifting of the sales limitation could be misinterpreted by the Soviet Union. I therefore felt that my decision should be made only when it was clear that the Soviets and other nations would not mistakenly think it indicated a weakening of our position.
>
> I have determined that our position now cannot be mistaken: The United States, along with the vast majority of nations, has condemned and remains opposed to the Soviet occupation of Afghanistan and other aggressive acts around the world. We will react strongly to acts of aggression wherever they take place. There will never be a weakening of this resolve.[28]

Secretary of State Alexander Haig, the leading opponent of lifting the Soviet grain embargo, argued, rather weakly, that removal of the embargo did not represent a lessening of the Reagan Administration's hard line toward the Soviet Union. Haig, stating that only the United States had "the pivotal strength" to stop Soviet aggression, called the Soviet Union "the greatest danger to world peace." Despite it being well-known that he was firmly opposed to the action, Haig stated: "It is my policy to fully support the presidency." Substantial speculation surrounded Haig's influence with other high administration officials, and a possible firing or resignation from his position.[29]

Secretary of Agriculture John R. Block had, on the other hand, clearly scored a major administration victory in direct confrontation with Secretary Haig. Block had vigorously opposed the embargo as Illinois Director of Agriculture during the Carter Administration and had continued his opposition as Agriculture Secretary. Soviet agriculture officials were in the Department of Agriculture building discussing grain purchases even before the official announcement was made. In addition to overall support of a "free-market" agricultural trade policy, the upcoming extension of the U. S. - Soviet grain agreement, a deadline on the administration's farm bill, and a predicted record wheat crop, all combined to strengthen Block's resolve to have the embargo removed.[30]

One senior administration official admitted that although the decision to end the grain embargo was probably the Reagan Administration's single most important foreign policy decision to date, "it was taken strictly, and I emphasize strictly, for domestic reasons." Administration officials stated that Reagan had never really changed his position on opposition to the embargo, and that it was necessary to act in order to secure votes in Congress for the administration's crucial economic package and farm legislation then under consideration. This position was supported by the fact that the administration did not move to lift controls on technology exports, nor on the prohibition against high-level political contacts between the United States and the Soviet Union.[31] The administration made no effort to attempt to seek any political or economic concessions from the Soviets in return, lending further credibility to the idea that the measure was taken for strictly domestic political reasons.

Chapter Notes

The Soviet Grain Embargo, January 4, 1980 - April 24, 1981

1. Terence Smith, "Carter Embargoes Technology to Soviet: Limits Fishing Privileges and Sale of Grain in Response to Aggression in Afghanistan," The New York Times, January 5, 1980.

2. Ibid.

3. Bernard Gwertzman, "U. S. Warns of New Responses to Soviet Over Afghanistan," The New York Times, January 6, 1980.

4. Ibid.

5. Ibid.

6. Ibid.

7. Seth S. King, "Halt Won't Affect Soviet Grain Supply," The New York Times, January 5, 1980.

8. Transcript of President's Speech, The New York Times, January 5, 1980.

9. Ibid.

10. King, "Halt Won't Affect Supply."

11. "Grain Export Halt to Harm U. S. Farmers, Economy," Wall Street Journal, January 7, 1980.

12. Ibid.

13. Ibid.

14. Ibid.

15. Seth S. King, "With or Without the Soviets, Farmers Depend on Exports," The New York Times, January 13, 1980.

16. Steven Rattner, "Trade as a U. S. Weapon," The New York Times, January 8, 1980.

17. Ibid.

18. U. S. House of Representatives, Committee on Foreign Affairs, Subcommittee on Europe and the Middle East, East-West Relations in the Aftermath of Soviet Invasion of Afghanistan, Hearings, January 24 and 30, 1980 (Washington, D. C.: U. S. Government Printing Office, 1980), p. 28.

19. Ibid., p. 41.

20. U. S. Senate Committee on Agriculture, Nutrition, and Forestry, Embargo on Grain Sales to the Soviet Union, Hearing, January 22, 1980 (Washington, D. C.: U. S. Government Printing Office, 1980), p. 59.

21. Ibid., pp. 13-14.

22. Ibid., p. 19. The domestic economic impact of the grain embargo was apparently not a major factor in the decision. Carter unquestionably felt he had to "get tough" in this situation, almost irrespective of the economic consequences which might result. The need to take a hard line on the Soviet invasion of Afghanistan was felt all the more necessary because of continuing criticism of the way Carter had handled the lingering hostage situation in Iran. One aide quoted Carter as saying: "Because of the way that I've handled Iran, they think I don't have the guts to do anything. You're going to be amazed at how tough I'm going to be." When Carter made the decision to impose the grain embargo, Treasury Secretary G. William Miller was vacationing in the Bahamas, and was not called home. See "Grain Becomes a Weapon," Time, January 21, 1980.

23. Ibid., pp. 24-25. According to Secretary Bergland, USDA was not called in to examine the potential impact of suspending grain sales to the Soviets until January 2, 1980. Marshall Shulman testified that U. S. intelligence had picked up indications of a possible Soviet invasion of Afghanistan as early as September. See U. S. House, East-West Relations, esp. p. 91.

24. Lou Cannon and Lee Lescaze, "U. S. to Keep Embargo on Grain Sales," Washington Post, February 5, 1981.

25. Ibid.

26. Philip Geyelin, "Sack the Embargo?", Washington Post, January 9, 1981.

27. "Reagan Undecided on Lifting Grain Embargo," Washington Post, February 6, 1981.

28. Lee Lescaze, "Reagan Ends Ban on Selling Grain to Soviet Union," Washington Post, April 25, 1981.

29. John M. Goshko, "Hard Line Toward Soviet Unchanged, Haig Insists," Washington Post, April 25, 1981.

30. Ward Sinclair, "Block Savors a Personal Vindication," Washington Post, April 25, 1981.

31. Bernard Gwertzman, "Grain Action and Politiccs," The New York Times, April 25, 1981.

CHAPTER 8

SUMMARY AND CONCLUSIONS

The major purpose of this study has been to develop an understanding of the formulation and purpose of United States international food policy. Observers in the field have normally tended to examine the food policy area from either the perspective of the policy formulation process, or from the standpoint of the international food situation, i.e., how policy interacts with global structure and supply conditions. In particular, an impressive volume of literature has emerged concerning the question of whether or not America's overwhelming dominance of the international food system provides the United States with the capacity to use its agricultural resources as an instrument in its overall relations with other states.

The concept of food power is most often interpreted as the withholding by one state of its food exports from another state, as a means of compelling the target government to conform to some form of behavior, or in order to "punish" a state for unacceptable, or at least undesirable behavior. As the case studies in U. S. food policy demonstrate, however, the objectives of foreign agricultural policy are often far more numerous and complex.

Determining the Success or Failure of Food Policy

It is not intended to suggest, of course, that the conditions under which food policy activities take place - in the international system, and within the country at which a particular action is directed - have no bearing on understanding policy. The case studies in United States food policy do indicate, however, that an understanding of such conditions does not by itself allow us to predict the success or failure of food policy.

Opinions on America's capability to utilize its agricultural resources for general foreign policy purposes run the full spectrum, from the position that food power provides the key to restoring the United States to its former position of pre-eminence in the world, to the belief that its vast agricultural resources actually create a kind of dependency situation on the major importing countries. Such theories regarding the potential utility of American food power are generally based on consideration of one or more of three factors: the world food supply situation; the nature of other countries' dependence on U. S. agricultural exports, including their ability to adjust to a disruption in supply, and the potential impact a suspension might have; and the domestic constraints on American central decision-makers to manipulate U. S. agricultural resources.

Determination of the success or failure of United States food policy actions is dependent only in part on the first two factors outlined above. The status of the world food supply condition, a target country's degree of dependence on U. S. food exports, its ability to develop alternative sources of supply or make adjustments in its consumption plans, and the eventual consequences of a suspension of food shipments may partly determine judgments of the success or failure of policy. It is unlikely, however, that any meaningful evaluation of these factors can be made by observers for purposes of making policy decisions.

A condition of world food scarcity, for example, has widely been regarded as a fundamental pre-condition that must exist in order for the United States to be capable of utilizing its food resources for purposes of national power. The determination that the international food supply is in "shortage" or "surplus," however, has a different meaning for virtually every country with any degree of participation in the world food system. The 1972-73 world food crisis did not, in fact, affect the "world" at all. Those countries which could afford to purchase food on the international market at higher prices continued to do so throughout the crisis. A number of them increased the level of their imports significantly, most notably the Soviet Union and Japan. Under almost any world food supply situation, those countries that are willing and able to pay the market price of food will be able to satisfy all or at least a good part of their needs.

The poor countries of the developing world, on the other hand, are likely to be dependent on purchasing food under the concessional terms offered by the major food exporting countries, mainly the United States, under any circumstances. And while world supply conditions may affect the price of food to some extent, any downward movement in price also serves to make exported food more attractive to rich countries. When the price of food is lower on the international market, major food importers such as Japan, the Soviet Union, and Western Europe may increase their purchases in order to expand the size of their domestic reserve stocks, or perhaps to use as feed grain to increase the amount of meat available to their citizens. These poor countries may not be in a position to purchase food on the world market regardless of supply conditions. This observation was supported in the 1977 House of Representatives study, "Use of Food Resources for Diplomatic Purposes":

> . . . food-poor LDC's will continue to be constrained by their poverty to obtain their food where the food is made available. Thus, for food-poor LDC's, the issue is not whether there is a surplus or shortage in world markets, but whether there is concessional food available.[1]

Since 1972, most international grain transactions have taken place primarily for the purpose of "improving" the diets of people in the developed countries as a consequence of rising affluence. Historically this has meant the exporting of cash crops from the developing areas to the Western industrialized states - coffee, tea, cocoa, bananas, and sugar, to name a few. More recently, a growing volume of food transfers have been occurring among developed countries as the populations of Western Europe, Japan and more recently, the Soviet Union and Eastern Europe demand larger amounts of meat in their diets.[2] As all of these areas are capable of producing sufficient quantities of grain required by their populations for direct consumption, the major function of grain imports by the developed countries is for providing expanding amounts of animal feed.[3]

Since most agricultural exports from the developing areas are also imported by the wealthy countries, the situation that has existed for some time shows wealthy states as the recipients of the vast majority of international agricultural trade. This pattern of dependence upon the wealthy countries as markets for agricultural exports has important implications with regard to the ability of many developing countries to meet their own food requirements through domestic production. Thus factors of international trade combine with internal factors to further limit the capabilities of Third World countries to supply their own food needs by altering present patterns of dependence on a few grain exporting states.

The vulnerabilities of particular developing counties are not of course equal, and vary according to a number of factors, including their degree of dependence on food imports, and how significant those imports are to the national diet. (See Appendix B.) There also exist differences in dependence among importers based on the particular commodity on which a country depends on having available in adequate supply. According to figures offered by Cheryl Christensen, fewer developing countries are producers of wheat, than of either corn or rice. A significant number of developing countries produce at least a majority of their corn or rice requirements and some even export small amounts. Christensen argues, therefore, that the degree of a particular country's "sensitivity" is dependent upon the significance of imports in the national diet. For example, few countries which are completely dependent on wheat imports count wheat as a primary staple. There are, however, certain Middle Eastern countries which rely on wheat as a staple, and are in fact dependent on wheat imports for a significant percentage of their total food consumption. Such countries would include Lebanon, Libya, Jordan, Saudi Arabia, Algeria, Iraq, Morocco and Tunisia, where wheat imports account for between 26 percent and 60 percent of daily calories. Chile is similarly dependent on imported wheat for almost 50 percent of daily calories.[4]

Another highly significant factor in considering a particular country's vulnerability to manipulation of grain exports is the aspect of which sectors of society are likely to suffer the consequences of supply disruption. In general, if withholding exports results in less supply, prices of a particular commodity will rise and the poorest sectors of society will suffer. It is this factor which may be the real obstacle to influencing successfully another state's policies. That group, the poor, which is most likely to be affected by the withholding of food exports is not likely to be in a position to influence national policy in most circumstances. Christensen cites the failure of U. S. food power under such conditions in India:

> Taking into account the salience of imports for national diets and which groups are likely to bear the brunt of reduced supplies, it is possible to offer some speculations about sensitivity. India has been a large wheat importer. Yet these imports have provided a relatively small portion of domestic consumption (10%) of a relatively modest diet component (14% of daily calories). These figures are suggestive in light of the political inefficacy of American attempts to gain political leverage by manipulating food shipments to India. When such figures are combined with the knowledge that concessional wheat imports have provided primarily subsidized food for the poor and a shield against cyclical starvation, it is easier to understand that the impact of decreased availability has fallen primarily on the poor (many of whom did not live to exert political pressure in response to shortages); wealthier consumers were able to make dietary adjustments and substitutions. Attempting to exercise 'behavioral power' under such circumstances combines ineffectiveness with a disastrous punishment of the least influential (economically and politically) within the country.[5]

However, the example of the U. S. attempt to undermine Chile's Allende regime by exercising food power yielded quite different results. Even though Chile's food imports were much smaller than India's, the effects of reduced food imports were apparently felt by a broader spectrum of society, and thus resulted in political pressures on the regime.

Thus, the efficacy of attempts to influence national policies or the stability of particular regimes may depend ultimately upon the political and socio-economic structure of a target country. In those cases where a reduction of food imports, or the commanding of higher prices for food imports affects certain sectors of a society, the impact felt by a government will depend on the ability of such groups to exert political pressure in response to reduced amounts of food. Such judgments may be difficult to make in advance. It would seem that in many cases, the groups most affected by disruptions in food supply would be those groups who have little or no political leverage. It is, of course, highly unlikely that the rulers of a country would be directly affected by a cutback in food imports in terms of their own diets.

Furthermore, there may exist political circumstances in certain countries that would work against a particular regime from responding to any sort of coercion by a foreign power, particularly if the source is the United States. For example, while Iran under the late Shah may have been susceptible to American attempts to wield food power, this is not the case under the present leadership, which would likely use such an event to further mobilize their population against the United States. Many regimes would also go to great lengths to deny an inability to feed their own people for purposes of maintaining their own position of power. Thus, countries with the most serious nutritional needs may prove to be least susceptible to influence based on food exports.[6]

There is considerable difference in opinion as to whether or not United States food policy can be used to influence Soviet behavior in any significant way. The question was in no way settled by the experience of the 1980 grain embargo. Those who initially supported the embargo subsequently found evidence to prove its success, while those who opposed the action declared it a complete failure, arguing that it hurt the United States more than the Soviet Union.

At least since 1972, the Soviet agricultural plan has been dependent in a significant way on imports. The magnitude of the Soviet need means that a major portion of their imports must come from the United States. Under the present international food situation, which is not likely to change in the near future, the United States is the only grain producer with sufficient production capacity to meet Soviet import needs without a total disruption of existing trade relationships.

During the recent grain embargo, the Soviets were able to make up a significant portion of the amount denied to them by the United States. The 1980 embargo was, furthermore, only a partial embargo. The Soviets still received the eight million tons of grain they were entitled to purchase under the terms of the U. S.-Soviet Grain Agreement, plus additional amounts that were already in the pipeline. Most observers believed that further quantities of American-produced grain eventually found their way to the Soviet Union through other channels.

Perhaps the most surprising aspect of the embargo was the fact that the American grain industry did not suffer, and 1980 concluded as a record year for exports. This was at least partly due to certain trade shifts that occurred as a result of Russian efforts to replace the suspended American grain. Even though the Soviets were able to make up a considerable portion of the embargoed grain, they did fall short of their net requirements, and were forced to pay higher prices for the grain they obtained from such sources as Argentina. In addition, they had to pay for much of this grain with scarce hard currency, instead of being permitted to make their purchases on concessional terms from the United States.

Any sort of internal distress that might have resulted would not easily be identified outside the Soviet Union. The Soviet population does not, of course, have much opportunity to protest conditions of any sort, and there is no reason to expect a sudden mobilization of the citizenry against the regime over what may have resulted in less availability of meat. Nevertheless, this does not mean that the impact of the 1980 grain embargo was not significant. The present Soviet leadership has made a well-publicized commitment to increase the amount of meat in its citizens' diets, and leaders since Khrushchev have been keenly aware of the growing need to satisfy consumer demands in a system that has become widely exposed to Western society.

Professor Alec Nove, for example, believes it was "politically significant" that the Soviet leadership kept the population informed about the embargo, attempting to place the blame for meat shortages on the United States. (Professor Nove believes that severe meat shortages did occur in the Soviet Union in 1980, and cites "large numbers of towns where there was no meat at all, for weeks and weeks on end.")[7] In addition, the degree to which Soviet planners may have had to reallocate resources away from other areas of the economy as a result of having to make up for the suspension of U. S. grain may never be known. If, for example, the Soviets were compelled to shift resources away from the defense sector, the implications would be quite significant, but such a situation cannot be easily evaluated by outside observers.

Writing in 1976, William Schneider supported the position that United States food policy could be effectively utilized toward the

Soviet Union, due in particular to certain aspects of Soviet agriculture.[8] First, while the Soviet Union traditionally dealt with shortfalls in production by "belt-tightening" measures before 1972, this course of action appears to have become less available to the present leadership. During the 1972 shortfall in production, the Soviets chose to import unprecedented quantities of foreign grain, primarily to feed their livestock, rather than slaughter the herds as had been done in prior shortages. Again in 1980, when the Soviets were denied seventeen million tons of grain they had expected to receive from the United States, they took every measure available to make up the shortage, rather than slaughter livestock. As mentioned, the price for these measures appears to have been substantial.

By choosing to exercise the "leverage" available as a consequence of its position as a major supplier of Soviet grain imports, Schneider believes the United States can significantly affect the area of resource allocation. Due to the notoriously inefficient nature of Soviet agriculture, the availability of relatively cheap imported grain allows the Soviets to substitute inexpensive American exports for costly domestic products. Alternatively, if the Soviets are compelled to produce similar quantities of additional grain themselves, or pay higher prices for non-American exports in order to feed their livestock, they would be required to significantly reallocate resources, probably away from the industrial and military sector. Thus, Schneider argues, the impact of withholding U. S. grain exports from the Soviet Union is likely to be substantial:

> Major shortages of feed grain would require drastic shifts in the allocation of resources from other sectors of the economy to the agricultural sector, thereby affecting Soviet industrial and military potential. In this regard, the manipulation of

> agricultural exports to the Soviet Union is more likely to have a significant impact on resource allocation within the Soviet Union than withholding industrial technology.[9]

In any event, however, it is highly improbable that United States food policy activities of any sort will produce immediate, dramatic results of any kind. The problem is that expectations are often unrealistically high, and results are therefore destined to be disappointing. Such expectations are likely to be misdirected not only among the public, but also among congressmen, the news media, academics, as well as policy makers, at least among those not directly involved in the decision process.

Forms of Economic Power

Klaus Knorr outlines four distinct purposes to which national economic capabilities can be used to achieve international leverage, or advantage gained by one actor over another. The first purpose is "coercion," which he defines as the withholding of something of value by one actor in order that another actor comply with some specified behavior. Such coercion exercised successfully constitutes "behavioral power."[10]

A second purpose of national economic power is the extraction of profit, which is possible when a state enjoys monopoly control over a particular commodity which other states depend upon in some way. The seeking of profit, Knoor states, is fundamentally similar to coercive economic power in that both depend on a state commanding a monopoly position over a commodity, and both demand a price as a result of such a position. In the first case, the price is compliance with some sort of desired behavior, and in the second case, monetary profit. There is a distinction between these two forms, however, which is more significant than the nature of compliance alone. For while a high price intended to extract maximum profit generally affects all customers, coercion is more often directed at a specific customer in an attempt to achieve particular objectives.[11]

The third possible use of national economic capability is the case when one state attempts to directly affect another state's economic well-being in order to weaken it in some manner. This is the more "traditional" use of national economic power, which is generally classified as "economic warfare" in times of military conflict, but can also be exercised in time of peace. The fourth purpose cited by Knorr is when one country gives "economic valuables," or some form of economic aid to another country in order to acquire a degree of influence over it.[12]

Knorr correctly points out that the ability of one state to exercise economic coercive power over another state arises from a situation of economic interdependence between two states. Situations of interdependence rarely occur among equals, and in fact may always be asymmetrical. The possession of national economic power by a state, moreover, need not be exercised explicitly. Power arises from the structural condition between actors in the international system, and need not be articulated in order to be realized. Thus, understanding the effects of such a possession of economic power may be difficult to detect, in much the same manner as the usefulness of nuclear deterrence is. According to Knorr:

> To threaten to inflict economic deprivation is not the only way through which coercive economic power can become internationally effective. Power is also a matter of perception and anticipation. B may be careful not to adopt courses of action to which

> A might object, because he expects that if he does, A might well resort to a threat. If B takes such care, his freedom of choice is restricted, and A's power has become effective without having been brought into play. It is highly likely that power becomes effective more often through this silent mechanism than as a result of deliberate threats and their execution. We cannot verify this hypothesis statistically. While the execution of international threats is always a matter of record, and making threats is also frequently made public (always in cases involving important stakes), the cautious act of rejecting, let alone not even considering, policies that might lead to unwelcome trouble, requires no communication between actors, usually occurs in private and often remains unrecorded.[13]

Albert O. Hirshman offers a similar analysis of what he terms the "influence effect" of foreign trade"

> . . . commerce can become an alternative to war . . . by providing a method of coercion of its own in the relations between sovereign nations. Economic warfare can take the place of bombardments, economic pressure that of sabre rattling . . . .
>
> For the political or power implications of trade to exist and to make themselves felt, it is not essential that the state should exercise positive action, i.e., organize and direct trade centrally; the negative right of veto on trade with which <u>every</u> sovereign state is invested is quite sufficient.[14]

Any evaluation of the potential purpose and usefulness of United States foreign agricultural policy must, however, take into consideration the process of policy formulation, as well as external conditions. No judgment of the "success" or "failure" of an American international food policy action is possible without understanding the objectives of the various central decision-makers involved, and which sets of interests are being supported in any particular case.

The Process of U. S. Food Policy, 1972 - 1980

The 1972 Soviet grain transactions were without doubt the most pivotal event for United States food policy during the decade of the 1970s. Although the sales were an integral part of a deliberate strategy to transform the nature of American foreign agricultural policy, it is unlikely that anyone could have predicted the magnitude of their impact.

While the evidence seems to indicate considerable collusion between USDA and the large grain trading firms in managing the sales, there is nothing to indicate that the corporations played a decisive role in planning the transactions. Clearly, corporate agriculture benefited from the sales, as well as from the overall expansion of American commercial agricultural exports. But while the grain corporations had attempted to bring about changes in U. S. food export policy at least as early as 1968, when they attempted to have the 50/50 vessel requirement lifted, it was not until "national" objectives entered the picture that major policy changes actually occurred.

As America's position in the international economy declined, and the Nixon Administration began planning what would become its New Economic Policy, officials took a closer look at ways of expanding U. S. commercial food exports. Also around this time, national security officials recognized trade expansion with the Soviet Union as an important instrument in support of detente.

Although the United States did sell about three million tons of grain to the Soviets in 1971, it was not until the following year that a major policy change on agricultural trade was decided upon by the Nixon Administration, opening up the way for the massive 1972 sales. When the directive was issued to begin exploring means for increasing food exports to the Soviets, the policy was issued not by the Department of Agriculture, but by Nixon's chief foreign policy strategist, national security adviser Henry Kissinger.

Expansion of trade with the Soviets had by this time become a major instrument in implementing detente with the Soviets. In 1968, corporate lobbying efforts to expand trade with the Soviets had clearly failed. But by 1970, national economic objectives entered the picture, and were subsequently strengthened by "national interest" goals held at the very highest level of the U. S. government. Thus the major areas of interest in the foreign economic policy arena all combined in support of a single direction - increasing sales of grain to the Soviets.

The decision process leading up to the 1975 U. S.-Soviet Grain Agreement was totally unlike that which resulted in the 1972 grain transactions. While national security officials welcomed a long-term grain agreement as an opportunity to exercise diplomatic leverage, the Agriculture Department strongly opposed an agreement as being detrimental to the interests of corporate agriculture.

By this period, however, control over U. S. foreign agricultural policy was moving increasingly away from USDA, and toward foreign policy officials, in particular Henry Kissinger. Kissinger was at this time actively utilizing food policy in support of U. S. foreign policy objectives in Chile, Indochina, and the Middle East, as well as in support of detente with the Soviets. He had also devised a strategy for using the food issue as a means of swaying Third World opinion against OPEC, and for gaining oil concessions from the Soviets, both of which failed. As a result, USDA was almost completely excluded from the food policy process.

The major feature of the decision-making process for U. S. food assistance in the early 1970s was the rather dramatic transformation which had occurred regarding the two major sets of objectives on the issue. The Agriculture Department had undergone a sharp change in its mission, from being charged with disposing of surplus food and market development through concessional programs, to supporting a free-trade system and the maximization of commercical exports.

The Department of State, which had used food assistance for many years to support a variety of diplomatic and general foreign policy objectives all over the world, had turned to the U. S. food aid program to support the war in Indochina to the exclusion of almost all other uses. As a consequence, the situation of complementary objectives shared for many years by USDA, State, and AID ended, and the issue was transformed into one in which the objectives of each of the major sectors with a stake in food aid conflicted.

AID's humanitarian objectives were thus completely pushed aside, just at the point when the poor nations of the Third World had become most desperate for U. S. food assistance. The Department of Agriculture no longer had any need for the food assistance program, its disposal and market development objectives having been accomplished. Furthermore, food aid priorities did not fit well with USDA's recently acquired commercial goals. Thus by 1973 the U. S. food assistance program was for all purposes held entirely within the domain of the Department of State, although major decisions were clearly controlled by Henry Kissinger personally.

Despite Kissinger's blatant use of food aid funds to support the war effort in Indochina and the military regime in South Korea, it was, nevertheless, Kissinger's personal power which set in motion the activities that culminated in the 1972 World Food Conference in Rome. Kissinger had apparently developed real interest in addressing the world food problem, whether his motivation was humanitarian or based on a fear of global instability.

Preparation of the U. S. position for the WFC was marked by interagency conflict between foreign policy officials, who wished to pledge the U. S. to increased levels of humanitarian food assistance, and domestic economic policy officials, who wished to limit levels of U. S. food aid, fearing the impact on domestic food price inflation. USDA at first opposed increased levels of food assistance, seeking to retain maximum quantities of agricultural commodities for commercial export. This position changed, however, as Agriculture officials began to predict bumper 1974 crops, and consequently feared that a commercial supply glut might severely dampen prices. Further complicating the situation was Kissinger's plan to use the U. S. WFC speech to attempt to force OPEC to lower the price of oil.

Despite strong opposition by Treasury Department and other domestic economic policy officials, who were concerned that increased levels of food assistance would hurt both anti-inflation efforts and the U. S. balance of payments position, Kissinger was able to capitalize on his personal power position. He delivered a major policy address before the United Nations in April, 1974, in which he pledged greater levels of U. S. food aid. Despite announcement of this commitment, Kissinger was still unable to convince President Ford to request a larger PL 480 allocation.

Even though Kissinger had considerable support in the Congress for expanding the levels of food assistance, Ford's major concern remained with the domestic economy. Ford shared neither Kissinger's nor Nixon's penchant for foreign policy, and was at this time gearing up his newly introduced Whip Inflation Now (WIN) program. Kissinger's position had been further weakened by a severe summer drought in the U. S., pushing food production down and prices up. The administration wanted to decrease, not increase, the food assistance program.

Needless to say, all these conflicting stakes and participants led to a rather hectic, confusing decision-process during this period. Deliberations on food policy were occuring within at least three different groups: the President's Committee on Food, and more particularly its sub-group, the Food Deputies Group, chaired by Gary Seevers of the CEA; an interagency committee chaired by OMB; and the State Department. The first two groups were oriented primarily toward the domestic economic implications of food policy, while State, or more specifically, Kissinger, was involved in preparing the U. S. position for the World Food Conference. Kissinger's strategy was to exclude economic policy officials from participating in the development of the U. S. position, knowing that they would oppose any attempt to pledge increased food assistance, as he had done before the UN in April, and again in September.

When Treasury Secretary William Simon learned of Kissinger's plan shortly before the WFC was scheduled to convene, he quickly moved to prevent its success, and thereby defend his own priorities. Just one week prior to the day Kissinger was scheduled to deliver the opening speech, Simon convened a meeting of the Economic Policy Board, a high level group which excluded representatives from either State or Agriculture. Kissinger was overseas, and the group succeeded in persuading President Ford to adopt their position.

The maintenance of a large grain reserve, held by the United States but in effect a world reserve, was an integral part of USDA farm policy for many years. With the transformation from a concessional system to a commercial system, however, the holding of large stocks of grain, either publicly or privately, no longer fit in with the Department's objectives. Particularly with the implementation of agriculture's role in the NEP, the grain corporations and their USDA representatives had no further use for a system which sought to stabilize prices, and was not in keeping with a policy of maximizing exports.

The first major action taken to eliminate American-held reserve stocks of grain was the engineering of the 1972 Soviet grain sales. This was followed in 1973 by that year's Farm Bill, which contained provisions specifically designed to further deplete grain stocks. USDA's position on the reserve issue was already clearly established.

Henry Kissinger, meanwhile, had begun to get involved in food policy at this time, particularly with regard to developing the American position for the 1974 World Food Conference. As part of the same strategy to increase the levels of U. S. food assistance to the Third World, Kissinger also sought to commit the United States to participate with other countries in establishing a major international grain reserve system. He announced such a pledge in his April 1974 speech before the UN General Assembly, and repeated the theme in his address before the World Food Conference. In addition, he succeeded in inserting a similar pledge in President Ford's 1974 speech before the United Nations.

Thus Kissinger pursued a course of action to support U. S. involvement in an international grain reserve system similar to the one he followed with respect to his attempt to increase the level of U. S.

food assistance before the World Food Conference. Specifically, he proposed the idea of a cooperative multilateral food reserve system to consist of an much as 60 million tons of food over carryover levels then held, to be controlled by various cooperative measures.

In contrast to USDA's rather neutral stance on Kissinger's effort to increase food aid, Agriculture was very clearly opposed to an international reserve system. In keeping with his support of corporate agriculture's interests, Butz was in favor of no reserve at all, but was apparently willing to live with a modest reserve. He was vehemently opposed, however, to a government-held reserve plan, such as the one proposed by Kissinger.

Important elements of the foreign policy community, meaning officials from other than AID, had by this period developed something of an appreciation for some of the demands of the Third World. Again, Kissinger in particular apparently had begun to recognize the possible implications of the "North/South conflict," and the need for some sort of constructive response to growing demands by the poor countries. By early 1974, broad agreement had been reached among the various administration factions over an international reserve scheme in which the member states, rather than multilateral agencies, controlled supplies. By the spring, however, the Nixon Administration was rocked by the Watergate scandal, and certain key economic policy officials left office, and were replaced by new faces. Most significant in this regard were the departures of George Shultz and Peter Flanigan, the latter a key architect of the NEP and Director of the Council on International Economic Policy. Shultz and Flanigan had served as perhaps the two major economic policy officials able to counterbalance Henry Kissinger's political/strategic objectives. Combined with President Nixon's subsequent resignation, Kissinger thus found himself in a highly advantageous position on the food issue.

In November, he succeeded in representing the U. S. position at the World Food Conference as one clearly favoring a major international system, even though the issue was far from being resolved internally. Following the WFC, Kissinger was able to gain further control over the reserve issue by persuading President Ford to create a new interagency committee under the jurisdiction of the National Security Council to handle follow-up measures to the Conference. The International Food Review Group (IFRG) provided Kissinger and his staff with the opportunity to frame the issue in national security terms, and the forum with which to persuade economic policy officials of their position on their own turf.

In spite of these early successes by Kissinger, due in large part to a domestic political crisis, he was unable in the end to succeed in his position. One reason was probably the fact that the foreign policy bureaucracy had little experience or expertise in the international food reserve issue, having been largely the personal bailiwick of Kissinger and Assistant Secretary of State Thomas Enders. Even after becoming Secretary of State, Kissinger never really utilized the Department bureaucracy, which proved to be no match on the issue for USDA's experience and expertise. And despite Kissinger's attempts to frame the international food reserve issue as a national security matter, its economic and agricultural policy implications far outweighed its strategic significance, a fact that he could not alter.

In the policy of withholding food assistance to the Salvador Allende regime in Chile, food policy was utilized as one component of a comprehensive foreign economic policy strategy. As such, it would have been unlikely to see USDA play a major role in the policy process. Throughout the Nixon Administration, USDA was, on various occasions, shouldered aside by "state sector" decision-makers, in areas which might have been expected to fall within Agriculture's domain.

The election of Allende in 1970 brought to the forefront many of the traditional cold war cliches, which both Nixon and Kissinger apparently genuinely believed. Despite the fact that Allende was leader of the Chilean Socialist Party, and had opposed the Communists for many years, Nixon and Kissinger quickly labeled him a Communist, and raised in their speeches all of the old fears of a Communist takeover not only of Chile, but all of Latin America.

The President, Kissinger, the CIA, and the Departments of Treasury and State all played a major role in determining Chile policy. Due in large part to the nature of the issue, USDA played no role at all. And in direct contradiction with the 1972 Soviet grain sales, in the Chile case major corporate interests clearly supported a cut-off of food transactions. While food exports were not a major factor in U. S. foreign economic policy toward Chile in 1970, corporations such as ITT, Kennecot, and Anaconda had massive financial stakes in the country, and supported using any methods available to remove Allende from power.

The Treasury Department also emerged at this time to play an important role in foreign economic policy, one which it had held traditionally. Particularly before Kissinger assumed the post of Secretary of State, Treasury maintained a higher level of influence than

State in certain areas of foreign policy, particularly when there was a prominent economic component involved. This was also in part due to the personal influence held by the two Secretaries of the Treasury under Nixon, George Shultz and John Connally, and the lack of power by Secretary of State William Rodgers, which was in turn related to Henry Kissinger's enormous influence in foreign policy of all types.

This whole power relationship under the Nixon Administration's foreign economic policy structure was also related to priorities under the NEP, which emphasized the international economic dimension of foreign policy, and therefore gave Treasury a more important voice in foreign affairs. This was particularly relevant with respect to foreign policy toward the Third World. Prior to NEP, Treasury had had to share more of these policy areas with State. In expropriations policy, State had a fairly soft line, while Treasury's was harder, in keeping with its role as representative of U. S. corporate interests. The voice of business in U. S. foreign economic policy under the Nixon Administration was further bolstered by Nixon's creation of the Council on International Economic Policy.

In the case of the new U. S. policy on expropriations announced by President Nixon in January, 1972, Treasury's hard line had clearly emerged victorious over State's softer approach. This decision also, of course, reflected U. S. corporate objectives, the draft of the plan having been prepared by officials of the International Telephone and Telegraph Company.

The degree of threat perceived over Chile by the corporate and ideological sectors at the highest level of the U. S. government far outweighed any USDA interests. Even though a highly significant food policy decision was involved, USDA was content to sit back and observe.

In a sharply contrasting case, the imposition of an embargo on all soybean exports in 1973 was clearly made on purely domestic economic grounds. Specifically, the decision was made amid fears that American soybean stocks were in short supply, and would not be sufficient for domestic needs. The policy clearly fell within the jurisdiction of the Commerce Department, which did in fact handle the matter. The fact that information concerning the supply situation was grossly inaccurate was partly due to the fact that the decision-making arena was kept rather narrow, excluding both foreign policy and foreign economic policy officials.

The fact that USDA was unwilling to cooperate in the establishment of any measures that would act to control, or even monitor exports, also affected the eventual decision to impose an embargo. Agriculture had the experience and expertise to gather the sort of information that was necessary to make a sound decision, but their unwillingness to cooperate dropped the ball into Commerce's lap. Although Commerce should have had the capability to fulfill what was technically its responsibility, the agency had little experience in the area of monitoring agricultural exports. Commerce lacked the staffing, organizational resources, and expertise in the agricultural area. An Agriculture Department not in opposition to export controls might have made up for Commerce's shortcomings. Earl Butz clearly had no inclination to do so.

Ironically, Agriculture's participation, particularly with regard to the information-gathering process, probably would have served to prevent imposition of the embargo, which it opposed. Had Commerce been provided with accurate information there is little doubt that the action never would have been ordered. Although Nixon's political advisers who moved to support the embargo also acted to force the decision, they probably never would have become involved had the supply figures not falsely pointed to the need for an export suspension. Yet another factor was the tremendous loss of prestige USDA had suffered as a result of its perceived incompetence in the 1972 Soviet grain transaction. In order to clear itself of any complicity in the "great grain robbery," Agriculture officials had pleaded ignorance of the whole course of events with the Russians. Their whole alibi was based on claims of lack of information, failure to predict, etc., even though the evidence subsequently refuted such claims. As a result, USDA was hardly in a position to dispute Commerce's figures.

Even though the decision to impose the soybean embargo probably would not have been made if Commerce's information had been accurate, the composition of the decision-making group also acted to shape the outcome. The ad hoc group under Gary Seevers that controlled the decision consisted solely of economic policy officials drawn from the Cost of Living Council and the Council of Economic Advisers. The mission of these officials was essentially domestically-oriented, to the exclusion of foreign policy or foreign economic policy concerns. Had the decision been controlled by the formal economic policy decision-making unit from which the ad hoc group had been established, the Council on Economic Policy, foreign policy officials would have been included in the decision process. The purpose of the CEP was, after all, to coordinate overall economic policy, domestic and foreign. Foreign policy officials would have likely opposed the decision, or at the very least, slowed it down. Such a delay may have allowed the time to discover that the information they were acting on was false. Even the unlikely circumstance that the decision would have been the same if considered by the whole CEP, foreign policy officials would at least have taken steps to notify important foreign customers of the embargo, and would thereby have minimized its negative impact abroad.

There was a broad consensus among both domestic and foreign policy advisers in the Carter Administration on the necessity to demonstrate some form of American response to the Soviet invasion of Afghanistan. While no one appeared to be in favor of any activities that might risk leading to a military confrontation, there clearly existed pressure to do something. President Carter had already demonstrated a tremendous amount of patience in his handling of the Iranian hostage situation, to the point of appearing indecisive and, to some, impotent. Carter had conveyed a somewhat weak impression on other foreign policy occasions as well, such as the incident involving the Soviet brigade in Cuba. Such failures to take decisive action threatened the administration's prestige abroad as well as the president's political standing at home.

The Carter Administration took careful steps to give the appearance that the embargo decision was one which was uniformly supported within its ranks. Carter's foreign policy style had been under severe criticism in the press, particularly with respect to friction between Secretary of State Vance and national security adviser Brzezinski. Nevertheless, some level of disagreement did exist among high-level foreign policy officials over the wisdom of employing a grain embargo, and on economic warfare strategy in general. Brzezinski was a strong proponent of linkage theory, and believed that the withholding of trade could be a highly effective weapon against the Soviets. Brzezinski's position was reportedly supported by the Department of Defense. As would be expected, the Commerce Department was on record as being strongly opposed to the use of trade for political purposes. State Department officials, while less enthusiastic about the embargo than Brzezinski, officially supported the action.

Secretary of Agriculture Bob Bergland was apparently completely excluded from the embargo decision. Although the potential consequences for the American agricultural community were obviously enormous, and were not unappreciated, the issue was clearly framed in national security terms. And although the action was in fact expected to cost President Carter some farm bloc votes, there existed a larger imperative to demonstrate to all voters some kind of decisive action. Provisions for softening the impact of the embargo on farmers were included as an integral part of the package, and it was on this matter that Secretary Bergland was called in to participate.

Although Ronald Reagan had strongly criticized Carter's grain embargo during the 1980 presidential campaign, and had promised to lift it after he assumed office, the new Secretary of State Alexander Haig tried desperately to change the president's mind. Haig far more closely resembled Brzezinski in his hard line position toward the Soviets than he did his predecessor at State. Haig believed that lifting the embargo would send the wrong signal to Soviet leaders, and would be totally inconsistent with the overall hard line that the Reagan Administration wished to project. Haig also argued that the grain embargo was a success, that it was indeed hurting the Soviets.

Secretary of Agriculture John Block was, like Reagan, clearly on record as opposing the grain embargo, and was not swayed by arguments against lifing it. And while the embargo issue had clearly been framed as a national security question in the Carter Administration, Block succeeded in having the issue considered before the full Cabinet. Despite mounting evidence in support of the embargo's impact on the Soviets, combined with a new Soviet threat being posed in Poland, Reagan lifted the embargo three months after assuming office.

Unlike the decision to impose the embargo, and unlike most food policy decisions which can be made quickly and/or quietly, the issue of lifting the embargo had become highly visible, and the subject of public debate for some fifteen months. Despite the fact that 1980 had been a record year for U. S. agricultural exports, and many farmers had supported the embargo all along for patriotic reasons, most farmers continued to oppose the action. Forces in the Congress continued to voice their opposition, led by farm state legislators such as Senator Robert Dole of Kansas.

The Carter Administration's decision to impose the 1980 grain embargo was a relatively simple decision made with little dissent among the potentially competing factions in the food policy arena. Although National Security Adviser Brzezinski was a stronger supporter of the action than Secretary of State Vance, there appeared to be agreement that some decisive action was needed, and that the grain embargo did not involve great risk. The one high-level official who could have been expected to oppose the embargo, Secretary of Agriculture Bergland, was left out of the initial decision process, and only called in to assist in softening its impact on farmers. Carter's domestic political advisers also supported the embargo, and economic policy officials either chose not to oppose the action, or were never given the opportunity. Similarly, the Congress was informed only after the decision was already made.

The decision to lift the grain embargo was, on the other hand, far more complex. President Carter had initially succeeded in framing the embargo as a national security matter, and executing the decision swiftly and with a very narrow range of participation. In the fifteen months after the decision was announced, however, considerable debate occurred on the issue, and a broad spectrum of actors became involved. Farmer associations, farm state congressmen, and pro-free trade interests vigorously opposed the embargo. The group of actors which maintained support of the embargo, however, remained small, consisting almost entirely of "state" actors. The embargo issue had, in addition, become a hot topic of debate in the presidential election campaign. Even though the embargo maintained support among many American citizens, the public had no real vital stake in the issue, and were therefore not an important factor after the election.

Although international developments - reports that the embargo was hurting the Soviets, the Polish situation - should have acted to lend further support to the embargo, once the matter had slipped into becoming a domestic political issue, they simply had no effect on the decision. On the other hand, mounting pressure to reactivate food exports to the Soviet Union, the upcoming extension of the U. S.-Soviet grain agreement, a predicted record wheat harvest, and an upcoming deadline on the administration's farm bill were all domestic factors that had major impact on the decision. The decision to employ the embargo, although it enjoyed support from domestic political advisers, had been decided solely on foreign policy grounds. The decision to lift the embargo, although opposed by foreign policy officials, was decided purely on domestic grounds.

A State Interest Approach to Understanding United States Foreign Agricultural Policy

A major premise of this study has been that U. S. foreign agricultural policy could be best understood through utilization of a "state interest" analytical framework. According to this approach, the foreign policy objectives of certain central decision-makers, primarily represented by the White House and the Department of State, can be conceived interms of the general interest of the "state" in international politics. Such objectives can often be expected to conflict with the particularistic interests of groups within society, and, in a general sense, seek to support the maintenance of the nation's security and economic well-being.

An examination of the case studies in the United States foreign agricultural policy process confirms the fact that food policy is normally the product of two broad sets of interests, domestic economic and foreign policy. This situation was true historically beginning with the period shortly after the end of World War II, when the United States emerged to occupy its dominant position in the world food system, at least until 1972.

The most important domestic economic objectives were generally thought of as supporting the income of American farmers; protecting the interests of domestic consumers; and promoting overall growth of the economy. Foreign policy interests sought to utilize food mainly as an instrument of foreign assistance in support of a broad range of United States goals abroad, ranging from strategic goals, to ideological interests, to humanitarian concerns. United States food exports throughout this period were generally transferred to other counties through the principal concessional networks under terms set forth in the Food for Peace Act. The U. S. food aid program was highly successful in accomplishing its principal objectives: disposal of surplus food and market development; earning of foreign exchange; and support of various American objectives abroad. These domestic economic and foreign policy objectives were, for the most part, highly complementary with one another, and the food policy process was one marked by general cooperation.

Following the sudden, dramatic transformation of United States food policy in the period beginning in 1972, from one in which large volumes of food transfers occurred through foreign assistance, to one that was almost totally commercial, this situation changed. Major United States domestic economic and foreign policy objectives no longer appeared congruent, and U. S. food policy became an area of sharp conflict among groups representing these interests.

This change can be traced primarily to alterations which occurred with respect to the domestic economic side of food policy. With the commercialization of U. S. foreign agriculture, competing sets of interests formed in which one set of actors, corporate agribusiness, emerged as a highly influential sector with a consistent stake in a free market agricultural economy, marked by high, fluctuating prices and maximum volume of exports. The achievement of these objectives, however, was not at all times complementary with the achievement of American domestic economic or foreign policy objectives. Neither was the achievement of these objectives at all times consistent with the best interests of American farmers.

Although American agriculture had been undergoing a major transition for many years, marked by the disappearance of the small farmer and his replacement by corporate agribusiness, it was not until Earl Butz was appointed the second Secretary of Agriculture in the Nixon Administration that USDA re-focused its policy direction to reflect these changes. The small farmer in America was an historical institution long regarded as an underpinning of society, and constantly courted for his electoral power. This importance decreased as he declined in number, and was overshadowed in many areas by corporate agribusiness which, after Butz's appointment, was reflected institutionally in USDA.

The economic interests of the American farmer had traditionally been perceived as an integral component of the whole domestic economy. Thus, in the area of food policy, agricultural interests were in most cases complementary with domestic economic interests, and vice versa. Institutionally, this situation was reflected in shared objectives among USDA and economic policy officials. The domestic economic area was normally represented by the Department of the Treasury, and various high-level policy groups such as the Council on International Economic Policy, the roles of which varied under different administrations.

With the transition in the nature of USDA's interests after the appointment of Earl Butz, however, this situation changed. A USDA acting as representative of corporate agriculture and the international grain traders could no longer be expected to share objectives with U. S. economic policy officials. And while the instruments utilized by foreign policy officials in attempting to achieve their objectives changed in certain respects, the policies supported by this sector could still be most usefully understood in terms of the interests of the state.

Thus throughout the period between 1972 and at least 1976, United States foreign agricultural policy can be more usefully understood as being the product of three, rather than two, competing sets of interests: "agricultural" policy; domestic economic policy; and foreign policy. However, an important distinction should be made in that domestic economic policy officials were not normally in a position to initiate actions in food policy, but could be expected to defend their area of interest. Agricultural policy officials had some limited ability, and could best achieve their interests by maintaining jurisdiction over policy. Foreign policy officials were by far the most active, dynamic sector in food policy, and when they did choose to intervene in an issue, were in most cases able to achieve some or all of their objectives.

United States food policy decision-making does resemble in many important ways the process described by the bureaucratic politics model. United States foreign agricultural policy is often a highly complicated process involving large numbers of actors with differing stakes in the outcome. Food policy potentially involves sets of interests representing almost every sector of the political process: military/-strategic; diplomatic; humanitarian; foreign economic; domestic economic; and of course, domestic political. Each of these sectors has its institutional representatives in the United States government. And the position of each of these sets of interests, as well as the eventual policy decision, is to a significant extent a product of both "bureaucratic politics" and "organizational process." Furthermore, the case studies in U. S. food policy have clearly demonstrated the importance of organizational jurisdiction in determining the final outcome. Thus, United States food policy decision-making does resemble in many important respects the process described by the so-called bureaucratic politics model.

Although the bureaucratic politics model does contribute a good deal to our understanding of U. S. foreign economic policy in general, including food policy, it does not in the end fully explain events. In other words, the bureaucratic politics model, as well as the perspective from which it descended, the political process paradigm, offers a necessary, but not sufficient explanation of the process of U. S. food policy-making.

The theoretical heritage of the political process paradigm, which played an instrumental role in diffusing the image of the rational actor approach in international political analysis, should not be overlooked. By focusing on the process of foreign policy formulation, the political process perspective correctly pointed to the fundamental role played by organizations, participants within organizations and, most importantly, the realization that like domestic policy, the formulation of foreign policy is political.

But the underlying contention of more recent bureaucratic politics approaches that organizational structure and bureaucratic routine are the primary, even sole determinants of the content of policy, has limited utility. The consequence of adopting such a perspective has been to obscure fundamental insights into understanding the objectives of foreign policy, thus losing a valuable feature of the formerly prominent realist approach.

The realist paradigm, which emerged during the 1950s as the major framework in American political science for studying the behavior of states in international politics, contributed significantly to the specialized study of foreign policy issues. While the systemic level of analysis provided a comprehensiveness which permitted analysis of the international system as a whole, its requirement of postulating a high degree of uniformity among national actors shifted focus away from the decision-making unit, and oversimplified the political process.

Largely in reaction to this shortcoming of the realist paradigm, in the early 1960s a group of writers began the first attempt to apply a political, policy-process perspective to foreign policy-making.[15] The political process paradigm played an important role in diffusing the image of the rational actor model in international politics. By focusing on the process of foreign policy formulation, the political process perspective correctly pointed to the fundamental role played by organizations, participants within organizations, and, most importantly, the realization that like domestic policy, the formulation of foreign policy was political. Analysts began to look beyond rational calculation of the national interest by a unified central decision-making unit by recognizing that the manner in which decisions are made influences the policy which results from those decisions.

The political process model was subsequently refined by what Robert Art refers to as the "second wave" of theorists who make up what is known as the bureaucratic politics school.[16] According to this perspective, organizational structure and bureaucratic procedure did more than influence policy - they determined its content.

> Bureaucratic theorists imply that it is exceedingly difficult, if not impossible, for political leaders to control the organizational web which surrounds them. Important decisions result from numerous smaller actions taken at different levels in the bureaucracy by individuals who have partially incompatible national, bureaucratic, political, and personal objectives. They are not necessarily a reflection of the aims and values of high officials.[17]

The bureaucratic politics paradigm went too far, however, in its claim that bureaucratic behavior and organizational process were the primary determinants of policy. The consequences of adopting the perspective have acted to obscure fundamental insights into understanding United States foreign policy. Central decision-makers do possess images, often strong ones, about how they think the world should look, and objectives based on more than their organization's mission. These objectives are crucial in understanding the activities of policymakers, and go well beyond the contention that organizational position is the sole determinant of policy stance, or, "where you stand depends on where you sit."

Such images are especially significant in analyzing the objectives of central decision-makers in those sectors of the government which can be characterized as "state institutions" in the foreign policy process, most notably the White House and Department of State. Unlike bureaus such as the Department of Agriculture, which represents the interests of particular societal groups, the "organizational essence" of state institutions is the general interest of the state in international affairs.

As a consequence of its myopic focus on organizational process and bureaucratic inertia, the bureaucratic politics perspective is unable to accommodate a fundamental, essential element in foreign policy formulation - central decision-makers' perception of the state interest. Such support of the state interest may at times be complementary with the pursuit of other areas, such as the particularistic economic interests of American corporations. This was the case, for example, in the decision to arrange the 1972 grain transactions. At other times, however, advancement of the state interest by central state actors may be almost totally incompatible with any of the other areas of interest at stake. This appears to have been the case in the Carter Administration's decision to impose the 1980 grain embargo against the Soviet Union.

The role of state actors is also special in another way, that the bureaucratic politics model also fails to account for. This is with respect to the special role of the president, who may elect to intervene in the policy process and direct how a particular issue is "framed," for example, as a security matter or agricultural policy question, thereby determining the organizational jurisdiction of the issue. Such an action is likely, although not in every case, to be crucial in determining the outcome of policy.

Presidential intervention in the policy process in this respect is in turn dependent on yet another important factor, the level of decision. As Francis E. Rourke stated in the preface to Bureaucracy and Foreign Policy:

> The political stakes of presidents and their personal perspectives on foreign affairs are more decisive than any other factors shaping the character of American involvement abroad. The conduct of foreign policy under President Nixon and the dominant position of his special assistant for national security affairs, Henry Kissinger, provide dramatic evidence of this fact on the contemporary scene.[18]

In other words, in those cases where the president perceives an interest and intervenes in the policy process, he has the ability to excise a decision from "standard operating procedures." A personal representative of the president may possess a similar ability, either on behalf of his own power, or as a consequence of being able to convince the president of his position. Most often the president or a personal representative will intervene in order to support a set of objectives associated with the interests of the state.

Edward K. Hamilton, in his summary report on "Case Studies in U. S. Foreign Economic Policy, 1965-1974," offers a scheme for classifying issues of U. S. foreign economic policy according to the level of decision they are generally perceived to require. He outlines six groupings:

a. Those which the President chooses to monitor personally. "These are usually not discrete decisions but a steady flow of events and choices in which the President is a fairly continuous participant."

b. Those in which the President is requested or chooses to make a single large decision, but has not acted as a continuous participant.

c. Those in which a direct representative of the President - either a person, agency, or group acting in his name - participates in or makes a decision.

d. Those decided in an "interagency context" by bureaucratic process.

e. Those which are decided wholly within the agency in which they arise.

f. "Those which never arise at all - initiatives unconceived or unimplemented, events unperceived or unreported, or policies unexamined."[19]

The "issue profile" of what is traditionally regarded as foreign economic policy is exceptional with regard to its "sister areas" in that it ordinarily is not perceived to require direct presidential intervention (categories a and b); neither can it normally be monitored and resolved at the agency or interagency level (categories e and d). In most issues of foreign economic policy, some lead or central person acting in the name of the president (category c) manages the decision process.

The need for a central figure or institution to manage the decision process of a foreign economic policy issue is largely a function of the number and diversity of individuals and institutions which possess a "legitimate" role in this area. According to Hamilton's analysis, all or most of the following actors are involved in most "serious issues:"

a. Those interested primarily with foreign matters. These are separated by "unmistakeable and often soundproof barriers" between those concerned with political, or "general" matters and those concerned with economic, or "specialized" issues.

b. Those primarily concerned with domestic matters. This distinction between political and economic is made here as well, but the distinction is not nearly as sharp.

c. Those primarily concerned with the military aspects of foreign policy.

d. Those responsible for the "overhead resource management processes of government" - essentially the budgetary process - and for macro-economic policy planning.

e. Those responsible for a particular policy event, or a particular ongoing policy sequence (e.g. tariff negotiations).

f. Relevant geographic specialists (e.g. in the State Department).

g. "Those responsible for knitting all of the strands - foreign and domestic, economic and political, functional and geographic - into a single body of national policy of reasonable internal consistency and maximally effective procedures to limit the damage when one or another specialized agency gets out of hand."[20]

The major organizational task is, of course, to be able to provide for the central management and coordination of the policy process (as described in g). According to Hamilton, the "convential wisdom," as specified by the existing statutes and Executive Orders which stipulate the current structure of the Executive Branch, is for this coordination to be provided by the Department of State. In most instances, however, the record of U. S. foreign economic policy suggests that the Department of State's record in filling this role is rather weak.

Nevertheless, the case studies in U. S. foreign agricultural policy clearly demonstrate that central state actors such as the president's national security advisor, or a high-level presidential economic assistant are most likely to perceive an issue in terms of the overall interest of the state. These officials are often able to intervene in a foreign economic policy issue, and manage the policy process so as to accomplish their objectives. They are, moreover, better equipped to provide a broader perspective than those sectors that represent the particularistic interests of groups in society, and thus better serve the national interest.

Due to the highly conflictual nature of the food policy process, however, the "capturing" of an issue by one sector or another generally means that the interests of competing sectors will not be sufficiently considered in the decision. When, for example, the national security sector has been able to succeed in controlling a food policy issue within its own jurisdiction, agricultural officials are not likely to be called upon to participate in the policy process, and their interests will probably not be considered. The same, of course, has been true in reverse. And because so many sets of interests are involved in the food policy process, competing sectors cannot really be expected to appreciate the potential implications of policy areas they are not directly involved in.

The cases examined in this study confirm the conflictual nature of the decision-making process in food policy, and sustain the hypothesis that this dynamic provides the key to understanding the objectives of policy. Routine, incremental decision-making in foreign agricultural policy normally occurs within the jurisdiction of the Department of Agriculture, where the objectives of "agricultural" interests may vary sharply, and have become increasingly diverse in recent years. Nonetheless, food policy objectives within USDA are not likely to coincide with those of the foreign policy sector.

In the event that a food policy issue comes to be perceived as having significance beyond the agricultural sector, however, control over the decision-making process may be wrested from the Department of

Agriculture. Should this occur, non-agricultural interests will generally be accorded priority. This "capturing" of an issue is most likely to occur when high level foreign policy actors believe that some sort of state objective can be accomplished by the use of agricultural resources, and succeed in framing the issue in terms of national security interests. In those circumstances where foreign policy officials intervene in a food policy issue in such a manner, they may or may not succeed in accomplishing their objectives.

In some cases Agricultural officials may simply abdicate their role to foreign policy officials, if they do not perceive a stake in an issue, of if they feel agricultural objectives will still be accomplished. In other circumstances domestic economic considerations may overide both agricultural and foreign policy priorities. And depending on a multitude of factors, including the personalities and positions of power of high level officials within an administration, Agricultural officials can succeed in resisting the efforts of foreign policy actors, and retain control of a particular issue.

The probability of foreign policy officials succeeding in gaining control over a food policy matter is highest when the issue can be framed in terms of the state interest, and considered within a short time frame, as a foreign policy "crisis". In such circumstances, foreign policy concerns are likely to be accorded the highest priority, irrespective of any opposing positions by domestic economic or agricultural officials. Furthermore, if the matter is considered as a crisis decision, domestic interests will not have the opportunity to mobilize any opposition. This was the case in the Carter Administration decision to impose the 1980 Soviet grain embargo.

Thus although foreign policy officials may attempt to utilize food policy to achieve their own objectives, they will not always succeed. In the case of the 1973 soybean embargo, the composition of the decision-making group excluded any foreign policy actors, and the officials involved failed to consider the foreign policy implications of their decision. On this occasion, foreign policy actors did not make any substantial attempt to control, or even seek inclusion in the decision structure.

In three other cases examined, foreign policy officials did attempt to control food policy decisions in order to achieve their own objectives, but were unsuccessful. During the preparation of the United States position for the 1974 World Food Conference in Rome, as well as the subsequent development of American policy on the proposed international food reserve system, foreign policy officials succeeded in gaining initial jurisdiction over the decision process. They did not, however, achieve their goals in the end. In the first instance, domestic economic policy officials intervened, and succeeded in partially defeating Kissinger's objective of an American commitment to provide increased food assistance. Agriculture officials no longer felt a need for the program, and remained essentially neutral.

In the case of the international reserve issue, Kissinger secured foreign policy leadership on the matter by creating a special ad hoc decision group controlled by foreign policy officials. Both domestic economic and agricultural officials perceived major stakes in the issue, however, and their combined opposition to State's objectives proved decisive in the end. After a lengthy policy debate, high level domestic policy officials relieved State of its leadership role and controlled the decision. Again, in the case of the Reagan Administration's decision to lift the Soviet grain embargo, domestic political and agricultural considerations prevailed over foreign policy concerns. This situation was, however, severely complicated by a change in administrations and, therefore, key actors.

In five other cases, foreign policy officials intervened in food policy issues, and succeeded in controlling the decisions, and achieving their objectives. In three of these cases - the 1972 Soviet grain transactions, U. S. food policy toward the Allende regime in Chile, and the utilization of food aid funds to support the war effort in Indochina - no substantial internal opposition existed to foreign policy official's objectives. In the case of the Soviet grain deal, USDA's objectives, as well as those of domestic economic officials, were wholly complementary with foreign policy goals. Neither group perceived any major stake in either of the other cases, and simply allowed State to control the issues for its own purposes.

In the case of the 1975 U. S.-Soviet Grain Agreement, foreign policy officials were able to exercise firm control over the negotiating process, and achieve their goals despite strong opposition by the Department of Agriculture. And finally, in the case of the decision to impose the 1980 Soviet grain embargo, foreign policy goals again predominated, and were again directly counter to the interests of Agriculture officials. In no other case were foreign policy officials as successful in framing the decision as a foreign policy matter. Despite important, well-recognized objections by USDA, as well as the obvious risks of antagonizing farmers in a presidential election year, the decision was made to employ the embargo.

USDA continues to dominate the foreign agricultural policy process on a day-to-day basis, as long as major stakes are not perceived by non-USDA actors. USDA remains the department with the expertise in the area, the access to vital information, and the connections within the agricultural community. Time and again, however, when food policy matters come to be perceived as major issues by non-USDA sectors, these actors move to control the issue by shifting the policy process out of USDA, and into their own jurisdictions. Once the particular issue fades, these officials have tended to turn their attention back to their primary concerns. Thus the food policy process is dispersed within the Federal Government, being controlled by agricultural officials on an everyday basis, and by foreign policy and/or economic policy officials in situations involving "major decisions."

There is a need for a more coherent foreign agricultural policy process, one that can provide a broad perspective, consider all the interests at stake, and one that possesses the experience and expertise to formulate policy in as rational a manner as possible. The logical locations for such a policy unit would be within the Department of State, which could bring the coherence and long-term perspective and responsiveness to food policy that is needed. Such a situation would only be possible, however, with strong support from the White House, which would have to refrain from intervening in major policy issues and removing them from State's jurisdiction, and with a cooperative Secretary of Agriculture with an appreciation for the overall implications of food policy.

Chapter Notes

1. U. S. Congress, House Committee on International Relations, Use of Food Resources for Diplomatic Purposes - An Examination of the Issues (Washington, D. C.: U. S. Government Printing Office, 1977), p. 5.

2. According to figures compiled by Cheryl Christensen, since the 1950s the developing countries have normally received less than one-third of total world grain exports. During 1972-73, the period of the U. S.-Soviet grain deal, the Soviet Union imported almost 90 percent of the total imports of all the developing countries combined. See Cheryl Christensen, "World Hunger: A Structural Approach," International Organization, Fall, 1978, pp. 755-758.

3. In fact, both Western Europe and Japan now produce a surplus of basic foodstuffs. While their production capabilities have improved, this also reflects a decreased demand for staple grains for direct consumption as dietary patterns changed.

4. Cheryl Christensen, "Food and National Security," in Klaus Knorr and Frank N. Trager (eds.), Economic Issues and National Security (Lawrence, Kansas: Regents Press, 1977), p. 303.

5. Ibid., pp. 303-304.

6. Joseph W. Willett and Sharon B. Webster, "Food Power: Food in International Politics," Political Aspects of World Food Problems (Kansas State University, July, 1978).

7. "Grain Embargo's Toll on U.S.S.R. and U. S. Will Rise with Time, "Wall Street Journal, January 24, 1980.

8. William Schneider, Food, Foreign Policy and Raw Materials Cartels (New York: Crane, Russak and Co., 1976), p. 59.

9. Ibid.

10. Klaus Knorr, "International Economic Leverage and its Uses," in Klaus Knorr and Frank N. Trager (eds.), Economic Issues and National Security, (Lawrence, Kansas: Regents Press, 1977), p. 99.

11. Ibid.

12. Ibid., pp. 99-100.

13. Ibid., pp. 102-103.

14. Albert O. Hirshman, National Power and the Structure of Foreign Trade (Berkeley, California: University of California Press, 1945), pp. 16-17.

15. Robert J. Art has identified the "first wave" of political process theorists as Roger Hilsman, Samuel Huntington, Richard Neustadt, and Warren Schilling. See Robert J. Art, "Bureaucratic Poltics and American Foreign Policy: A Critique," in John E. Endicott and Roy W. Stafford, Jr. (eds.), American Defense Policy, 4th edition (Baltimore: Johns Hopkins University Press, 1977).

16. Prominent bureaucratic politics theorists include Graham Allison and Morton Halperin.

17. Stephen D. Krasner, "Are Bureaucracies Important? (Or Allison Wonderland)," in Richard G. Head and Ervin J. Rokke, American Defense Policy, 3rd edition (Baltimore: Johns Hopkins University Press, 1973), p. 311.

18. Francis E. Rourke, Bureaucracy and Foreign Policy (Baltimore: Johns Hopkins University Press, 1972), p. vii.

19. Edward K. Hamilton, Introduction to the Summary Report, "Cases on a Decade of U. S. Foreign Economic Policy: 1965-1974," in Commission on the Organization of the Government for the Conduct of Foreign Policy, 7 volumes (Washington, D. C.: U. S. Government Printing Office, 1975).

20. Ibid., pp. 8-9.

SELECTED BIBLIOGRAPHY

Books

Adelman, Irma, and Morris, Cynthia Taft. Economic Growth and Social Equity in Developing Countries. Stanford, California: Stanford University Press, 1973.

Agribusiness Council. Agricultural Initiative in the Third World. Lexington, Massachusetts: Lexington Books, 1975.

Allison, Graham T. Essence of Decision. Boston: Little, Brown & Co., 1971.

Allison, Graham T., and Szanton, Peter. Remaking Foreign Policy: The Organizational Connection. New York: Basic Books, 1976.

Aronowitz, Stanley. Food, Shelter and the American Dream. New York: Seabury Press, 1974.

The Atlantic Council Working Group on the United States and the Developing Countries. The United States and the Developing Countries. Boulder, Colorado: Westview Press, 1971.

Aziz, Sartaj. Hunger, Politics, and Markets. New York: New York University Press, 1975.

Bahr, Howard J., et al., eds. Population, Resources, and the Future: Non-Malthusian Perspectives. Provo, Utah: Brigham Young University Press, 1972.

Baldwin, David A. Economic Development and American Foreign Policy: 1943-62. Chicago: University of Chicago Press, 1966.

Bard, Robert L. Food Aid and International Agricultural Trade. Lexington, Massachusetts: Lexington Books, 1972.

Bauer, Raymond A.; Pool, Ithiel De Sola; and Dexter, Lewis Anthony. American Business and Public Policy. Chicago: Aldine Publishing Co., 1963, 1972.

Bauer, Robert A., ed. The Interaction of Economics and Foreign Policy. Charlottesville: University Press of Virginia, 1975.

Beard, Charles A. The Idea of National Interest. New York: Macmillan, 1934.

Bell, John Fred. A History of Economic Thought. 2nd ed. New York: Ronald Press, 1967.

Bentley, Arthur F. The Process of Government. Chicago: University of Chicago Press, 1908.

Bhagwati, Jagdish N., ed. Economics and World Order. New York: Free Press, 1972.

Black, Lloyd D. The Strategy of Foreign Aid. Princeton, New Jersey: D. Van Nostrand Co., Inc., 1968.

Blake, David H., and Walters, Robert S. The Politics of Global Economic Relations. Englewood Cliffs, N. J.: Prentice-Hall, 1976.

Borgstrom, George. The Food and People Dilemma. Belmont, California: Duxbury Press, 1973.

Boulding, Kenneth E., and Mukerjee, Tapan., eds. Economic Imperialism: A Book of Readings. Ann Arbor, Michigan: University of Michigan Press, 1972.

Brandon, Donald. American Foreign Policy: Beyond Utopianism and Realism. New York: Appleton-Century-Crofts, 1966.

Brown, Lester R. By Bread Alone. New York: Praeger, 1974.

Brown, Lester R. World Without Borders. New York: Vintage Books, 1973.

Brown, Peter G., and Shue, Henry, eds. Food Policy. New York: Free Press, 1977.

Brown, Seyom. The Faces of Power. New York: Columbia University Press, 1968.

Brynes, Asher. We Give to Conquer. New York: W. W. Norton & Co., 1966.

Brzezinski, Zbigniew. Between Two Ages: America's Role in the Technetronic Era. New York: The Viking Press, 1970.

Buck, Phillip W. The Politics of Mercantilism. New York: Henry Holt & Co., 1942.

Bundy, William P., ed. The World Economic Crisis. New York: W. W. Norton, 1975.

Caldwell, Dan. Food Crises and World Poltics. Beverly Hills, California: Sage Publications, 1977.

Carr, Edward Hallett. The Twenty Years' Crisis, 1919-1939. 2nd ed. New York: Harper & Row, 1946.

Center for Science in the Public Interest. From the Ground Up: Building a Grass Roots Food Policy. Washington, D. C.: Center for Science in the Public Interest, 1976.

Central Intelligence Agency. Potential Implications of Trends in World Population, Food Production, and Climate. C.I.A., August, 1974.

Chomsky, Noam. American Power and the New Mandarins. New York: Vintage, 1969.

Clark, Colin. Starvation or Plenty? New York: Taplinger Publishing, 1970.

Clark, Colin. The Myth of Overpopulation. Melbourne, Australia: Advocate Press, 1973.

Clark, Paul G. American Aid for Development. New York: Praeger, 1972.

Cochrane, Willard W. Feast or Famine: The Uncertain World of Food and Agriculture and its Policy Implications for the United States. Washington, D. C.: National Planning Association, 1974.

Cochrane, Willard W. The World Food Problem. New York: Thomas Y. Crowell Co., 1969.

Cohen, Benjamin, ed. American Foreign Economic Policy. New York: Harper & Row, 1968.

Cohen, Stephen D. The Making of United States International Economic Policy. New York: Praeger, 1977.

Connelly, Philip, and Perlman, Robert. The Politics of Scarcity: Resource Conflicts in International Relations. London: Oxford University Press, 1975.

DeMarco, Susan, and Sechler, Susan. The Fields Have Turned Brown: Four Essays on World Hunger. Washington, D. C.: The Agribusiness Accountability Project, 1975.

Destler, I. M. Making Foreign Economic Policy. Washington, D. C.: Brookings Institution, 1980.

Destler, I. M. Presidents, Bureaucrats and Foreign Policy. Princeton, New Jersey: Princeton University Press, 1972.

DeVries, Egbert, and Richter-Altschaffer, J. H. World Food Crisis and Agricultural Trade Problems. Beverly Hills: Sage, 1974.

Dorner, Peter. Land Reform and Economic Development. Baltimore, Maryland: Penguin, 1972.

Eckholm, Erik P. Losing Ground: Environmental Stress and World Food Prospects. New York: W. W. Norton, 1976.

Eells, Richard. Global Corporations. New York: The Free Press, 1976.

Endicott, John E., and Stafford, Roy W., Jr., eds. American Defense Policy. 4th ed. Baltimore: Johns Hopkins University Press, 1977.

Freis, Herbert. Foreign Aid and Foreign Policy. New York: St. Martins Press, 1964.

Fox, Douglas M., ed. The Politics of U. S. Foreign Policy Making. Pacific Palisades, California: Goodyear Publishing, 1971.

George, Susan. How the Other Half Dies: The Real Reasons for World Hunger. Montclair, New Jersey: Allanheld, Osmun & Co., 1977.

Gilpin, Robert. U. S. Power and the Multinational Corporation. New York: Basic Books, 1975.

Goldman, Marshall I. Detente and Dollars: Doing Business with the Soviets. New York: Basic Books, 1975.

Goulet, Denis. The Cruel Choice: A New Concept in the Theory of Development. New York: Atheneum, 1971.

Guither, Harold D. The Food Lobbyists. Lexington, Massachusetts: D. C. Heath and Co., 1980.

Halperin, Morton H. Bureaucratic Politics and Foreign Policy. Washington, D. C.: The Brookings Institution, 1974.

Hamilton, Martha M. The Great American Grain Robbery and Other Stories. Washington, D. C.: Agribusiness Accountability Project, 1972.

Hansen, Roger D. A "New International Economic Order"? An Outline for a Constructive U. S. Response. Washington, D. C.: Overseas Development Council, July, 1975.

Hansen, Roger D., and the Staff of the Overseas Development Council. The U. S. and World Development: Agenda for Action 1976. New York: Praeger, 1976.

Hartz, Louis. The Liberal Tradition in America. New York: Harcourt, Brace and World, Inc., 1955.

Hayter, Teresa. Aid as Imperialism. Middlesex, England: Penguin Books, 1971.

Head, Richard G., and Rokke, Ervin J., eds. American Defense Policy. 3rd ed. Baltimore: Johns Hopkins University Press, 1973.

Heckscher, Eli F. Mercantilism. 2 vols. London: George Allen & Unwin, Ltd., 1935.

Helleiner, G. K. A World Divided: The Less Developed Countries in the International Economy. Cambridge University Press, 1976.

Helleiner, G. K. International Trade and Economic Development. Middlesex, England: Penguin, 1972.

Hightower, Jim. Eat Your Heart Out: Food Profiteering in America. New York: Crown Publishers, Inc., 1975.

Hilsman, Roger. To Move a Nation. New York: Delta, 1967.

Hirschman, Albert O. National Power and the Structure of Foreign Trade. Berkeley, California: University of California Press, 1945.

Holsti, K. J. International Politics: A Framework for Analysis. Englewood Cliffs, New Jersey: Prentice-Hall, 1972.

Horowitz, David. Hemispheres North and South. Baltimore: Johns Hopkins University Press, 1966.

Howe, James W., and the Staff of the Overseas Development Council. The U. S. and World Development: Agenda for Action 1975. New York: Praeger, 1975.

Iowa State University Center for Agricultural and Rural Development, ed. U. S. Trade Policy and Agricultural Exports. Ames, Iowa: Iowa State University Press, 1973.

Irish, Marion, and Frank, Elke. U. S. Foreign Policy. New York: Harcourt Brace Jovanovich, Inc., 1975.

Johnson, Gale D. World Food Problems and Prospects. Washington, D. C.: American Enterpise Institute for Public Policy Research, June 1975.

Johnson, Gale D., and Schnittker, John A., eds. U. S. Agriculture in a World Context. New York: Praeger, 1974.

Johnson, Robert H. Managing Interdependence: Restructuring the U. S. Government. Washington, D. C.: Overseas Development Council, February, 1977.

Jones, David. Food and Interdependence: The Effect of Food and Agricultural Policies of Developed Countries on the Food Problems of Developing Countries. London: Overseas Development Institute, 1976.

Junker, Louis, ed. The Political Economy of Food and Energy. Ann Arbor, Michigan: Graduate School of Business Administration, University of Michigan, 1977.

Kahn, Herman. The Next 200 Years. New York: William Morrow & Co., 1976.

Keohane, Robert O., and Nye, Joseph S. Power and Interdependence. Boston: Little, Brown, 1978.

Kinkleberger, Charles. Power and Money. New York: Basic Books, 1970.

Knorr, Klaus. Power and Wealth. New York: Basic Books, 1973.

Knorr, Klaus, and Trager, Frank N. Economic Issues and National Security. Lawrence, Kansas: Regents Press of Kansas, 1977.

Kolko, Gabriel. The Roots of American Foreign Policy. Boston: Beacon Press, 1969.

Kolko, Joyce and Gabriel. The Limits of Power. New York: Harper & Row, 1972.

Krasner, Stephen D. Defending the National Interest. Princeton, New Jersey: Princeton University Press, 1978.

LaFeber, Walter. America, Russia, and the Cold War, 1945-1975. 3rd ed. New York: John Wiley and Sons, 1976.

Lappe, Frances Moore, and Collins, Joseph. Food First: Beyond the Myth of Scarcity. Boston: Houghton Mifflin, 1977.

Lerza, Catherine, and Jacobson, Michael, eds. Food for People Not for Profit. New York: Ballantine, 1975.

Lewis, W. Arthur. The Evolution of the International Economic Order. Princeton, New Jersey: Princeton University Press, 1978.

Lindblom, Charles E. Politics and Markets. New York: Basic Books, 1977.

Lowi, Theodore J. The End of Liberalism: The Second Republic of the United States. 2nd ed. New York: W. W. Norton & Co., 1979.

Lucas, George R. Jr., and Ogletree, Thomas W., eds. Lifeboat Ethics: The Moral Dilemmas of World Hunger. New York: Harper & Row, 1976.

McConnell, Grant. Private Power and American Democracy. New York: Alfred A. Knopf, 1967.

McCune, Wesley. The Farm Bloc. Garden City, N. Y.: Doubleday, Doran and Co., Inc., 1943.

Marx, Herbert L., Jr., ed. The World Food Crisis. New York: H. W. Wilson Co., 1975.

Manson, Edward S. Foreign Aid and Foreign Policy. New York: Harper & Row, 1964.

Mass, Bonnie. Population Target: The Political Economy of Population Control in Latin America. Ontario: Charter's Publishing Co., 1976.

Mellor, John W. The Economics of Agricultural Development. Ithaca, New York: Cornell University Press, 1966.

Mesarovic, Mihajlo, and Pestel, Eduard. Mankind at the Turning Point - The Second Report to the Club of Rome. New York: E. P. Dutton & Co., 1974.

Minchinton, Walter E., ed. Mercantilism: System or Expediency? Heath & Co., 1969.

Morgan, Dan. Merchants of Grain. New York: Viking Press, 1979.

Morgenthau, Hans J. A New Foreign Policy for the United States. New York: Praeger, 1969.

Morgenthau, Hans J. Politics Among Nations. 5th ed. revised. New York: Alfred A. Knopf, 1978.

Myrdal, Gunnar. The Challenge of World Poverty. New York: Vintage, 1970.

Nathan, James A., and Oliver, James K. United States Foreign Policy and World Order. Boston: Little, Brown, 1976.

Nelson, Joan M. Aid, Influence, and Foreign Policy. New York: Macmillan, 1968.

Neustadt, Richard E. Presidential Power. New York: Mentor, 1964.

Packenham, Robert A. Liberal America and the Third World. Princeton, New Jersey: Princeton University Press, 1973.

Paddock, William and Elizabeth. We Don't Know How: An Independent Audit of What They Call Success in Foreign Assistance. Ames, Iowa: Iowa State University Press, 1973.

Parenti, Michael. The Anti-Communist Impulse. New York: Random House, 1969.

Payer, Cheryl. The Debt Trap: The IMF and the Third World. New York: Monthly Review Press, 1974.

Petras, James F., and LaPorte, Robert, Jr. Cultivating Revolution: The United States and Agrarian Reform in Latin America. New York: Random House, 1971.

Petras, James F., and Morely, Morris. The United States and Chile: Imperialism and the Overthrow of the Allende Government. New York: Monthly Review Press, 1975.

Pirages, Dennis. Global Ecopolitics. North Scituate, Massachusetts: Duxbury Press, 1978.

Political Aspects of World Food Problems. Kansas State University, July 1978.

Pontecorvo, Giulio, ed. The Management of Food Policy. New York: Arno Press, 1976.

Rhodes, Robert I., ed. Imperialism and Underdevelopment: A Reader. New York: Monthly Review Press, 1970.

Robbins, William. The American Food Scandal: Why You Can't Eat Well on What You Earn. New York: William Morrow, 1974.

Roll, Eric. A History of Economic Thought. Englewood Cliffs, New Jersey: Prentice-Hall, Inc., 1956.

Rosenau, James N., ed. International Politics and Foreign Policy. New York: Free Press, 1969.

Rosenau, James N. The Scientific Study of Foreign Policy. New York: Free Press, 1971.

Rourke, Francis E. Bureaucracy and Foreign Policy. Baltimore, Johns Hopkins University Press, 1972.

Schattschneider, E. E. Politics, Pressures, and the Tariff. Hamden, Connecticut: Archon Books, 1963. Originally published in 1935.

Schneider, William. Food, Foreign Policy, and Raw Materials Cartels. National Strategy Information Center, Inc. New York: Crane, Russak & Co., Inc., 1976.

Schumacher, E. F. Small is Beautiful. New York: Harper & Row, 1973.

Schumann, Franz. The Logic of World Power. New York: Pantheon Books, 1974.

Simon, Paul and Arthur. The Politics of World Hunger. New York: Harper's Magazine Press, 1973.

Singer, Hans W., and Ansari, Javed A. Rich and Poor Countries. Baltimore: Johns Hopkins University Press, 1977.

Snyder, Richard C.; Bruck, H. W.; and Sapin, Burton. Foreign Policy Decision-Making. New York: The Free Press of Glencoe, 1962.

Sorokin, Elena P. Hunger as a Factor in Human Affairs. Gainesville, Florida: University of Florida, 1975.

Spanier, John. American Foreign Policy Since World War II. 6th ed. New York: Praeger, 1974.

Spanier, John, and Uslaner, Eric M. How American Foreign Policy is Made. New York: Praeger, 1974.

Spanier, John, and Uslaner, Eric M. How American Foreign Policy is Made. 2nd ed. New York: Praeger, 1978.

Spero, Joan Edelman. The Politics of International Economic Relations. New York: St. Martin's Press, 1977.

Stanley, Robert G. Food for Peace. New York: Gordon and Breach, 1973.

Stillman, Edmund, and Pfaff, William. Power and Impotence. New York: Vintage Books, 1966.

Talbott, Ross. The World Food Problem and U. S. Food Politics and Policies: 1972-1976. Ames, Iowa: Iowa State University Press, 1977.

Tantner, Raymond, and Ullman, Richard H., eds. Theory and Policy in International Relations. Princeton, New Jersey: Princeton University Press, 1972.

The New York Times. Food and Population: The World in Crisis. New York: New York Times Co., 1975.

The New York Times. Give Us This Day. . . New York: Arno Press, 1975.

Thompson, W. Scott, ed. The Third World: Premises of U. S. Policy. San Francisco: Institute for Contemporary Studies, 1978.

Thorp, Willard L. The Reality of Foreign Aid. New York: Praeger, 1971.

Toma, Peter A. The Politics of Food for Peace. Tucson, Arizona: University of Arizona Press, 1967.

Truman, David B. The Governmental Process. New York: Alfred A. Knopf, 1951.

Tucker, Robert W., and Watts, William, eds. Beyond Containment; U. S. Foreign Policy in Transition. Washington, D. C.: Potomac Associates, 1973.

Tudge, Colin. The Famine Business. New York: St. Martin's Press, 1977.

Turner, Louis. Multinational Companies and the Third World. New York: Hill and Wang, 1973.

United Nations. Progress in Land Reform. 6th report. New York: United Nations, 1976.

Vicker, Ray. This Hungry World. New York: Charles Scribner's Sons, 1975.

Viner, Jacob. The Long View and the Short. Glencoe, Illinois: Free Press, 1958.

Wallerstein, Mitchel B. Food For War - Food For Peace. Cambridge, Massachusetts: MIT Press, 1980.

Ward, Barbara; Runnalls, J. D.; and D'Anjou, L. The Widening Gap: Development in the 1970's. New York: Columbia University Press, 1971.

Warley, T. K. Agriculture in an Interdependent World: U. S. and Canadian Perspectives. Montreal: Canadian-American Committee, 1977.

Waterlow, Charlotte. Superpowers and Victims: The Outlook for World Community. Englewood Cliffs, New Jersey: Prentice-Hall, 1974.

Weissman, Steve. The Trojan Horse: A Radical Look at Foreign Aid. San Francisco: Ramparts Press, 1974.

Wilber, Charles K., ed. The Political Economy of Development and Underdevelopment. New York: Random House, 1973.

Wilson, Charles. Mercantilism. London: Wyman & Sons, Ltd., 1958.

Wilson, James Q. Political Organizations. New York: Basic Books, 1973.

Wolfers, Arnold. Discord and Collaboration. Baltimore: Johns Hopkins University Press, 1962.

Wright, Quincy. The Study of International Relations. New York: Appleton-Century-Crofts, Inc., 1955.

Journals and Newspapers

Andriole, Stephen J. "Resource Scarcity and Foreign Policy: Implications for Research and Analysis." World Affairs, Summer 1976.

Balz, Daniel. "Agriculture Report: Politics of Food Aid." National Journal.

Balz, Daniel J. "Exporting Food Monopolies." The Progressive, January 1975.

Balz, Daniel. "State-Agriculture Feud Delays Grain Reserve System." National Journal, June 29, 1975.

Barraclough, Geoffrey. "The Great World Crisis I." The New York Review of Books, January 23, 1975.

Barraclough, Geoffrey. "The Haves and the Have Nots." The New York Review of Books, May 13, 1976.

Barraclough, Geoffrey. "Waiting for the New Order." The New York Review of Books, October 26, 1978.

Barraclough, Geoffrey. "Wealth and Power: The Politics of Food and Oil." The New York Review of Books, August 7, 1975.

Bergsten, C. Fred; Keohane, Robert O.; and Nye, Joseph F., Jr. "International Economics and International Politics." International Organization, Winter 1975, Vol. 29, No. 1.

Bergsten, C. Fred, and Krause, Lawrence B., eds. "World Politics and International Economics." Special edition. International Organization, Winter 1975, Vol. 29, No. 1.

Bryson, Reid A., and Ross, John E. "Climatic Variation and Implications for World Food Production." World Development, 1977, Vol. 5, Nos. 5-7.

Brzezinski, Zbigniew. "U. S. Foreign Policy: The Search for Focus." Foreign Affairs, July 1973.

Bundy, William P. "Elements of Power." Foreign Affairs.

Cardoso, Fernando Henrique. "The Consumption of Dependency Theory in the United States." Latin American Research Review, 1977, Vol. XII, No. 3.

Chilcote, Ronald H. "A Question of Dependency." Latin American Research Review, 1978, Vol. XIII, No. 2.

Christensen, Cheryl. "World Hunger: A Structural Approach." International Organization, Fall 1978.

Cleaver, Harry M., Jr., "The Contradictions of the Green Revolution." Monthly Review, June 1972.

Conner, Cliff. "Hunger: U. S. Agribusiness and World Famine." International Socialist Review, September 1974.

Cooper, Richard N. "Trade Policy is Foreign Policy." Foreign Policy, 9, Winter 1972-73.

Crosson, Pierre R. "Institutional Obstacles to Expansion of World Food Production." Science, May 1975, Vol. 188, No. 4188.

Destler, I. M. "United States Food Policy 1972-1976: Reconciling Domestic and International Objectives." International Organization, Fall 1978.

Diwan, Romesh. "Projections of World Demand for and Supply of Feedgrains: An Attempt at Methodological Evaluation." World Development, 1977, Vo. 5, Nos. 5-7.

Ehrlich, Paul R., and Ehrlich, Anne H. "The World Food Problem: No Room for Complacency." Social Science Quarterly, September 1976.

Fagen, Richard R. "The United States and Chile: Roots and Branches." Foreign Affairs. 53, January 1975.

Farnsworth, Elizabeth; Feinberg, Richard; and Leeson, Eric. "Facing the Blockade." NACLA's Latin America and Empire Report, 7, January 1973.

Foster-Carter, Aidan. "From Rostow to Gunder Frank: Conflicting Paradigms in the Analysis of Underdevelopment." World Development, 1976, Vol. 4, No. 3.

Frank, Andre Gunder. "Dependence is Dead, Long Live Dependence and the Class Struggle: An Answer to Critics." World Development, 1977, Vol. 5, No. 4.

Frank, Andre Gunder. "Economic Crisis, Third World and 1984." World Development, 1976, Vol. 4, Nos. 10/11.

Frank, Charles R., Jr., and Baird, Mary. "Foreign Aid: Its Speckled Past and Future Prospects." International Organization, Winter 1975, Vol. 29, No. 1.

Gelb, Leslie H., and Lake, Anthony. "Washington Dateline: Less Food, More Politics." Foreign Policy, 17, Winter 1974-75.

Geyelin, Philip. "Sack the Embargo?" Washington Post, January 9, 1981.

Gilpin, Robert. "Three Models of the Future." International Organization, Winter 1975, Vol. 29, No. 1.

Gold, David; Lo, Clarence; and Wright, Eric. "Recent Developments in Marxist Theories of the Capitalist State." Monthly Review, October 1975, Vol. 27, No. 5.

Goshko, John M. "Hard Line Toward Soviet Unchanged, Haig Insists." Washington Post, April 25, 1981.

"Grain Becomes a Weapon." Time, January 21, 1980.

"Grain Embargo's Toll on U.S.S.R. and U. S. Will Rise with Time." Wall Street Journal, January 24, 1980.

"Grain Export Halt to Harm U. S. Farmers, Economy." Wall Street Journal, January 7, 1980.

Grant, James P. "Food, Fertilizer, and the New Global Politics of Resource Scarcity." Annals of the American Academy of Political and Social Science, July 1975.

Gwertzman, Bernard. "Grain Action and Politics." The New York Times, April 25, 1981.

Gwertzman, Bernard. "U. S. Warns of New Responses to Soviet over Afghanistan." The New York Times, January 6, 1980.

Hansen, Michael K., and Risch, Stephen J. "Food and Agriculture in China." Science for the People, May/June 1979.

Hoffman, Stanley. "The Uses of American Power." Foreign Affairs.

Hopkins, Raymond F., and Puchala, Donald J. "Perspectives on the International Relations of Food." International Organization, Fall 1978.

Josling, Tim. "Grain Reserves and Government Agriculture Policies." World Development, 1977, Vol. 5, Nos. 5-7.

Katzenstein, Peter J. "Introduction: Domestic and International Forces and Strategies of Foreign Economic Policy." International Organization, Autumn 1977.

King, Seth S. "Halt Won't Affect Soviet Grain Supply." The New York Times, January 5, 1980.

King, Seth S. "With or Without the Soviets, Farmers Depend on Exports." The New York Times, January 8, 1980.

Kissinger, Henry A. "A New National Partnership." Department of State Bulletin, 72, February 17, 1975.

Kissinger, Henry A. "The Challenge of Interdependence." Department of State Bulletin, 70, May 6, 1974.

Krasner, Stephen D. "State Power and the Structure of International Trade." World Politics, April 1976.

Krasner, Stephen D. "U. S. Commercial and Monetary Policy: Unravelling the Paradox of External Strength and Internal Weakness." International Organization, Autumn 1977.

Lescaze, Lee. "Reagan Ends Ban on Selling Grain to Soviet Union." Washington Post, April 25, 1981.

Meltzer, Ronald, et. al. "United States Trade and Foreign Policy." Current History, May/June 1979.

NACLA. "U. S. Grain Arsenal." Latin America and Empire Report, 9, October 1975.

Nau, Henry R. "The Diplomacy of World Food: Goals, Capabilities, Issues and Arenas." International Organization, Fall 1978.

Paarlberg, Robert L. "Shifting and Sharing Adjustment Burdens: The Role of the Industrial Food Importing Nations." International Organization, Fall 1978.

Perlmutter, Amos. "The Presidential Political Center and Foreign Policy: A Critique of the Revisionist and Bureaucratic-Political Orientations." World Politics, October 1974.

Poleman, Thomas T. "World Food: A Perspective." Science, May 1975, Vol. 188, No. 4188.

Rattner, Steven. "Trade as a U. S. Weapon." The New York Times, January 8, 1980.

"Reagan Undecided on Lifting Grain Embargo." Washington Post, February 6, 1981.

Robbins, William. "Cargill: Big Grain Risk Taker." The New York Times, January 9, 1980.

Rosenfeld, Stephen S. "The Politics of Food." Foreign Policy, Spring 1974.

Rothschild, Emma. "Food Politics." Foreign Affairs, 54, January 1976.

Runge, Carlisle Ford. "American Agricultural Assistance and the New International Economic Order." World Development, 1977, Vol. 5, No. 8.

Sanderson, Fred H. "The Great Food Fumble." Science, May 1975, Vol. 188, No. 4188.

Schertz, Lyle P. "World Food: Prices and the Poor." Foreign Affairs, April 1974.

Schertz, Lyle P., and Berntson, Byron. "The New Politics of Food." World Development, 1977, Vol. 5, Nos. 5-7.

Seevers, Gary L. "Food Markets and Their Regulation." International Organization, Fall 1978.

Sinclair, Ward. "Block Savors a Personal Vindication." Washington Post, April 25, 1981.

Smith, Terence. "Carter Embargoes Technology to Soviet: Limits Fishing Privileges and Sale of Grain in Response to Aggression in Afghansistan." The New York Times, January 5, 1980.

Sondermann, Fred A. "The Conquest of the National Interest." Orbis, Spring 1977.

Sundquist, W. B. "The Changing Structure of U. S. Agriculture: Implications for World Trade." World Development, 1977, Vol. 5 Nos. 5-7.

Thompson, Louis M. "Weather Variability, Climatic Change, and Grain Production." Science, May 1975, Vol. 188, No. 4188.

Valenzuela, J. Samuel, and Valenzuela, Arturo. "Modernization and Dependency: Alternative Perspectives in the Study of Latin American Underdevelopment." Comparative Politics, July 1978.

Vermeer, Donald E. "Food, Farming, and the Future: The Role of Traditional Agriculture in the Developing Areas of the World." Social Science Quarterly, September 1976.

Wallensteen, Peter. "Scarce Goods as Political Weapons: The Case of Food." Journal of Peace Research, 1976, Vo. XIII, No. 4.

Walsh, John. "U. S. Agribusiness and Agricultural Trends." Science, May 1975.

Walters, Harry. "Difficult Issues Underlying Food Problems." Science, May 1975.

"Who's Helping Who?: A Radical Reassessment of Foreign Aid." New Internationalist, January 1976, No. 35.

Wilson, Mark. "Food as a Weapon." Science for the People, May/June 1979.

Zwerdling, Daniel. "The Food Monopolies." The Progressive, January 1975.

U. S. Government Documents

Commission on the Organization of the Government for the Conduct of Foreign Policy. 7 vols. Washington, D. C.: U. S. Government Printing Office, 1975.

U. S. Congress, House, Committee on Agriculture, Sale of Wheat to Russia. Hearings, September 14, 18, and 19, 1972. Washington, D. C.: U. S. Government Printing Office, 1972.

U. S. Congress, House, Committee on Foreign Affairs. Subcommittee on Europe and the Middle East, East-West Relations in the Aftermath of the Soviet Invasion of Afghanistan. Hearings, January 24 and 30, 1980. Washington, D. C.: U. S. Government Printing Office, 1980.

U. S. Congress, House, Subcommittee on Inter-American Affairs. United States and Chile During the Allende Years, 1970-1973. Briefing, March 6, 1973. Washington, D. C.: U. S. Government Printing Office, 1973.

U. S. Congress, House, Committee on International Relations, Use of U. S. Food Resources for Diplomatic Purposes - An Examination of the Issues. Washington, D. C.: U. S. Government Printing Office, 1977.

U. S. Congress, Senate, Committee on Agriculture and Forestry, Subcommittee on Foreign Agricultural Policy. Hunger and Diplomacy: A Perspective on the U. S. Role at the World Food Conference. Washington, D. C.: U. S. Government Printing Office, 1975.

U. S. Congress, Senate, Committee on Agriculture, Nutrition and Forestry. Embargo on Grain Sales to the Soviet Union, Hearing, January 22, 1980. Washington, D. C.: U. S. Government Printing Office, 1980.

U. S. Congress, Senate, Committee on Foreign Relations. Multinational Corporations and United States Foreign Policy. Hearings, March and April, 1973. Washington, D. C.: U. S. Government Printing Office, 1973.

U. S. Congress, Senate, Committee on Government Operations. Russian and Grain Transactions. Hearings, July 20, 23, and 24, 1973. Washington, D. C.: U. S. Government Printing Office, 1973.

U. S. Congress, Senate, Committee on Government Operations. Russian Grain Transactions. Report No. 93-1003. Washington, D. C.: U. S. Government Printing Office, 1974.

U. S. General Accounting Office. Russian Wheat Sales and Weaknesses in Agriculture's Management of Wheat Export Subsidy Program. Report to Congress. Washington, D. C.: GAO, 1973.

U. S. General Accounting Office. U. S. Actions Needed to Cope with Commodity Shortages. Washington, D. C.: GAO, 1973.

United States International Economic Policy in an Interdependent World. Report to the President, submitted by the Commission on International Trade and Investment Policy. Report and papers in 2 vols. Washington, D. C.: U. S. Government Printing Office, 1975. Also known as "Williams Commission."

Weekly Compilation of Presidential Documents. June 14, 1971. Washington, D. C.: General Services Administration.

Weekly Compilation of Presidential Documents. January 1, 1980. Washington, D. C.: General Services Administration.

Unpublished Materials

Frundt, Henry John. "American Agribusiness and U. S. Foreign Agricultural Policy." Unpublished Ph.D. dissertation, Rutgers University, 1975.

Weber, Thomas William. "Agricultural Exports and Decision-Making in American Foreign Policy." Unpublished Ph.D. dissertation, University of Virginia, May 1977.

Wilkinson, Kenneth P. "Malthus and/or Marx?" Unpublished monograph, 1977.

APPENDIX A

TABLES ON INTERNATIONAL FOOD TRADE

Table 1 - U. S. Exports as a Percent of World Exports

	wheat	grain*
1961/62	41.7	43.2
1963/64	39.6	42.2
1965/66	38.2	45.3
1967/68	38.0	41.8
1969/70	30.0	34.7
1970/71	35.2	35.2
1971/72	30.1	37.1
1972/73	44.0	50.0
1973/74	44.2	47.4
1974/75	41.1	46.4

*Grains include wheat and feedgrains
Source: Cheryl Christensen, "Food and National Security," in Klaus Knorr and Frank N. Trager (eds.), Economic Issues and National Security, (Lawrence, Kansas: Regents Press, 1977), p. 294.

Table 2 - U. S. Production as a Percent of World Production

	wheat	grain
1961/62	14.8	25.8
1963/64	13.1	25.8
1965/66	11.7	25.4
1967/68	13.9	25.8
1969/70	12.7	24.2
1970/71	11.7	22.0
1971/72	12.7	25.4
1972/73	12.4	27.4
1973/74	12.6	23.8
1974/75	13.9	20.7

Source: Christensen, "Food and National Security," p. 295.

Table 3 - Average Wheat Export Prices, 1968-1976 (U. S.)

Year	Price (dollar per bushel)
1968	1.69
1969	1.67
1970	1.74
1971	1.69
1972	1.86
1973	3.55
1974	5.16
1975	4.79
1976	3.98

Source: Raymond F. Hopkins and Donald J. Puchala, "Perspectives on the International Relations of Food," International Organization 32 (Summer, 1978), p. 584.

Table 4 - Net Exports and Imports of Grains (excluding rice)

	Millions of Metric Tons (+ = exports; - = imports)			
	1970-71	1971-72	1972-73	1973-74
United States	+38.3	+41.3	+70.7	+74.9
Canada	+15.8	+18.4	+18.9	+14.4
Australia	+12.3	+10.7	+ 5.1	+ 7.2
Western Europe	-27.6	-19.0	-18.7	-21.6
Japan	-15.3	-15.2	-17.5	-19.4
USSR	+ 7.5	- 1.0	-19.6	- 4.6
Eastern Europe	- 7.9	- 8.9	- 8.0	- 5.2
China	- 3.7	- 3.3	- 6.1	- 7.7
Developing Countries	-15.4	-26.9	-23.2	-30.3

Source: Fred H. Sanderson, "The Great Food Fumble," Science 188, May 9, 1975, p. 504.

Table 5 - World Grain Trade, 1934-1976

Region	1934-38	1948-52	1960	1970	1972-73	1976
			(million metric tons)			
North America	5	23	39	56	89	94
Latin America	9	1	0	4	-3	-3
Western Europe	-24	-22	-25	-30	-18	-17
E. Europe & USSR	5	--	0	0	-26	-27
Africa	1	0	-2	-5	-1	-10
Asia*	2	-6	-17	-37	-38	-47
Australia & N. Z.	3	3	6	12	7	8

Note: Positive numbers indicate net exports; negative numbers net imports
*Includes Japan and Asian Communist countries.
Source: Hopkins and Puchala, "Perspectives on Food," p. 586.

Table 6 - Wheat Reserve Stocks of Major Exporting Countries

(million metric tons)

Year	Exporting countries beginning stocks
Avg. 1960/61 through 1970/71	42.7
1971/72	44.4
1972/73	41.4
1973/74	22.7
1974/75	19.8
1975/76	22.3

Source: Hopkins and Puchala, "Perspectives on Food," p. 585.

Table 7 - Food Shipments Under PL 480 (dollar sales)

Commodity	1960	1965	1970	1972	1973	1974
			(thousands of metric tons)			
Wheat and products	8,199	13,705	5,765	4,615	2,517	1,005
Milk (dried)	8	42	18	19	2	0
Rice	453	561	884	813	987	620
Corn, sorghum	787	728	1,078	1,217	1,289	454
Vegetable oils	339	364	240	193	107	148

Source: Lester Brown, _By Bread Alone_, (New York: Praeger, 1974), p. 65.

Table 8 - Food Shipments Under Title II (grants)

Commodity	1960	1965	1970	1972	1973	1974
			(thousands of metric tons)			
Wheat and products	979	1,473	1,464	1,614	1,649	718
Milk (dried)	147	199	133	115	26	0
Rice	89	0	7	248	33	0
Corn, oats, sorghum and products	359	498	330	257	246	379
Blended food products	0	0	149	266	195	182
Vegetable oils	0	106	81	187	111	53

Source: Brown, By Bread Alone, p. 66.

Table 9 - U. S. Agricultural Exports by Major Categories

(millions of dollars)

Year	PL 480	AID	Commercial	Total
1970	1,056	12	5,650	6,718
1971	1,023	56	6,674	7,753
1972	1,058	66	6,922	8,046
1973	954	84	11,864	12,902
1974	867	76	20,350	21,293
1975*	1,093	123	20,368	21,584

*Preliminary
Source: Lawrence Witt, "Food Aid, Commercial Exports, and the Balance of Payments, in Peter G. Borwn and Henry Shue, Food Policy, (New York: Free Press, 1977), p. 83.

Table 10 - Shares of World Grain Production and Consumption*

(percentages)

	1960/61-62/63		1969/70-71/72		1973/74-76/77	
	prod.	cons.	prod.	cons.	prod.	cons.
Developed countries (U. S., Canada, EEC, Japan, Oceana)	37.3	35.4	36.1	33.5	35.7	31.2
Centrally planned countries (USSR, China)	34.6	35.0	35.8	36.9	36.4	38.2
Developing countries (Brazil, Argentina, S. Asia, N. Africa/Mideast	27.4	28.6	27.5	28.8	27.2	29.7

*Includes wheat, milled rice, and coarse grains.
Source: Henry Nau, "The Diplomacy of World Food: Goals, Capabilities, Issues and Arenas," International Organization 32, (Summer, 1978), p. 779.

Table 11 - Shares of World Grain Exports*

(percentages)

	1948/49 -51/52	1958/59 -61/62	1968/69 -71/72	1972/73 -73/74
North America	63.2	62.6	52.0	64.8
Western Europe	3.7	9.9	19.5	18.0
Centrally planned	6.9	4.3	2.4	1.0
Oceana	9.6	8.6	10.1	5.3
South America	9.8	9.5	10.3	6.5

*Includes wheat and wheat four, rye, barley, oats, maize, sorghum and millets. Does not include trade among centrally planned countries.
Source: Henry Nau, "Diplomacy of Food," p. 781

APPENDIX B

WHY POOR COUNTRIES CANNOT EASILY INCREASE THEIR OWN FOOD PRODUCTION

A major problem with food embargoes, some observers point out, is that the targets of food cut-offs can be expected to respond by increasing their own productive capacity within a few years time. Such analyses ignore, however, the economic and political dynamics of food production, particularly in the developing countries of the Third World.

Understanding why the food-dependent countries of the Third World appear unable to produce sufficient quantities of food to feed their own populations involves examining the social, political, and economic structures which determine the process of agricultural production. In many countries of the Third World, the structures of agricultural production which exist today have their roots in the colonial legacies of these areas. Under colonialism, many of the counties of Asia, Africa and Latin America were organized as "agricultural establishments" for the primary purpose of producing and supplying food to "the larger community to which they belong."[1]

The major alteration fostered upon the colonized areas by the colonial powers was the introduction of an agricultural system for the production of cash crops which replaced their traditional diversified agriculture. It is essential to recognize that the single crop economies which exist today in many Third World countries are not the result of natural development based on comparative advantage, but rather can be traced directly to the colonial period. In many instances abandonment of traditional food crops in favor of export crop production was economically, or even physically forced on local populations by the European colonizers. In other cases colonial administrations or foreign corporations simply took control of the land for the purpose of growing cash crops. The example of what occurred in Africa is typical of the alterations in agricultural production which took place under the colonizing powers in many parts of Asia and Latin America as well:

> Prior to European intervention, Africans practiced a diversified agriculture that included the introduction of new food plants of Asian or American origin. But colonial rule simplified this diversified production to single cash crops - often to the exclusion of staple foods - and in the process sowed the seeds of famine. Rice farmming once had been common in Gambia. But with colonial rule so much of the best land was taken over by peanuts (grown for the European market) that rice had to be imported to counter the mounting prospect of famine. Northern Ghana, once famous for its yams and other foodstuffs, was forced to concentrate solely on cocoa. Most of the Gold Coast thus became dependent on cocoa. Liberia was turned into a virtual plantation subsidiary of Firestone Tire and Rubber. Food production in Dahomey and southeast Nigeria was all but abandoned in favor of palm oil; Tanganyika (now Tanzania) was forced to focus on sisal and Uganda on cotton.[2]

Although colonialism has ended in all areas of the Third World, the process of agricultural production developed during the colonial period remains a crucial element in understanding why most developing countries appear unable to produce basic foodstuffs in quantities sufficient to feed their own populations. The infrastructures developed under colonialism to facilitate the production and export of cash crops for European markets remain as fundamental impediments to agricultural production for domestic needs in many developing areas.[3] The processes which over hundreds of years acted to destroy the capabilities of traditional agricultural societies to feed their own populations and consequently become dependent on imported food cannot be easily reversed, even if ruling groups in these countries should make the unlikely decision to do so.

Since independence, the colonial administrators who formerly controlled agricultural production in the Third World have in many countries been replaced by new groups composed of large landowners, local ruling elites, and Western multinational corporations.[4] These groups control most of the land in the countries of Asia, Africa, and Latin America, and decide which crops, if any, will be produced. Consequently, vast areas of the best land in all three continents are devoted to growing cash crops - coffee, tea, cocoa, cotton, rubber, flowers, sugar, bananas and peanuts - rather than basic foodstuffs to feed the local population. As long as profitable markets exist for these export crops, and comparable markets are not present for basic foodstuffs for local consumption, it is unlikely that ruling groups will sacrifice their economic interests in favor of growing increased amounts of food.

Futhermore, cash crops not only occupy vast areas of land that might otherwise be devoted to the production of food for domestic needs, they often occupy the best land and command vital inputs as well. The introduction of the Green Revolution to the Third World during the 1960s further compounded this situation.[5] Green Revolution technology requires expensive inputs - petrochemical fertilizer, insecticides, and complex agricultural machinery - which can only be afforded by large landowners. Since small farmers are not normally able to afford the enormous expenses for such inputs, they are often forced to sell out to larger landholders, which results in fewer and fewer people controlling agricultural production. As increasing numbers of small farmers are displaced from the land, more and more land is devoted to producing cash crops for export. Thus, in many areas of the Third World, production of basic foodstuffs for domestic consumption has been decreasing in recent years not only on a per capita basis, but in real terms as well.

In light of these circumstances, it is highly naive to suggest that potential targets of food embargoes, particularly in the Third World, can circumvent the efforts of a termination of food imports by simply increasing their domestic agricultural productivity.

Furthermore, the global food system which exists today is highly institutionalized and cannot be easily dissolved.[6] This system places additional international constraints on the ability of food deficit countries in the Third World to improve their domestic food production. This global food system, which governs both commercial channels and a concessional network of food "aid" has a clear structure which divides participants into food surplus countries and food deficit countries or, quite simply, sellers and buyers.

The present global food system under which a small group of grain exporting countries are depended upon to produce sufficient quantities of basic foodstuffs to feed enormous numbers of people in all corners of the world did not come about haphazardly. Most important in the present context is the understanding that this system of international dependence is not a "natural" circumstance resulting from either superior environment or technological conditions in certain areas of the world.[7]

The primary force which has shaped the global food system since the early 1950s has been United States foreign agricultural policies. These policies have in many instances served to reinforce historical patterns of agricultural production in the Third World, particularly with respect to specialized single crop or "few crop" economies. American programs which succeeded in disposing of surplus production through export promotion, credit grants, concessional arrangements and the development and expansion of overseas markets for U. S. agricultural products had important consequences for agricultural development in many developing areas.[8] This structure remains as a significant source of the apparent inability of Third World countries to produce basic foodstuffs for their own needs.

One reason American food export policies acted to constrain domestic food production was that in many Third World states, concessional food imports were often sold by governments in regular domestic markets, where these commodities directly competed with domestically produced food. This practice resulted in depressed prices which were often below the costs of domestic production, creating a situation which forced large numbers of farmers out of business, thus depressing food production. The development of foreign markets for cheap food exports by these means was of course an explicit objective of Public Law 480. While the consequences of these food transfers may not have been anticipated by American policymakers, they were, nevertheless, quite serious.

Another area of U. S. policy which had the effect of depressing food production in the Third World concerned American protectionist trade policies. The United States and, later, Western Europe employed trade barriers designed to protect certain agricultural commodities from foreign imports on the domestic market. This acted to prevent foreign producers from entering developed countries' markets in basic foodstuffs areas. Since indigenous markets remained limited in terms of effective demand, Third World producers had little choice but to continue producing "traditional" export products, meaning cash crops. Attempts to expand production of these traditional commodities so as to increase foreign exchange earnings often resulted in placing developing states in competition with one another, with detrimental effects. Furthermore, massive American (and to a lesser extent, Canadian and Australian) concessional programs prevented entry into basic foodstuffs markets withi the Third World itself.[9]

Appendix B - Notes

1. From a quotation of John Stuart Mill cited in Frances Moore Lappe, Food First: Beyond the Myth of Scarcity (Boston: Houghton Mifflin, 1977), p. 77

2. Ibid., p. 78.

3. Susan George points out that until the beginning of the Second World War, Africa, Asia and Latin America all were grain exporting areas. See George, How the Other Half Dies: The Real Reasons for World Hunger (Montclair, New Jersey: Allanheld, Osmun & Co., 1977).

4. According to figures provided by Susan George, in South America, 17 percent of landowners control 90 percent of the land; in Asia, 20 percent of landowners control 60 percent of the land; and in Africa, 25 percent of landowners control approximately 95 percent of the land. See Ibid., p. 14.

5. For further information on the Green Revolution, see Ibid., and Lappe, Food First.

6. Raymond F. Hopkins and Donald Puchala in a 1978 article developed the concept of "global food regime." Raymond F. Hopkins and Donald J. Puchala, "Perspectives on the International Relations of Food," International Organization, Fall 1978.

7. See discussions of the concept of "comparative advantage" in George, How the Other Half Dies, and Lappe, Food First.

8. Christensen, Cheryl, "World Hunger: A Structural Approach," International Organization, Fall 1978, p. 761.

9. Ibid.

For Product Safety Concerns and Information please contact our EU representative GPSR@taylorandfrancis.com
Taylor & Francis Verlag GmbH, Kaufingerstraße 24, 80331 München, Germany

www.ingramcontent.com/pod-product-compliance
Lightning Source LLC
Chambersburg PA
CBHW070548270326
41926CB00013B/2245

* 9 7 8 0 3 6 7 2 7 5 9 4 5 *